METHODS IN MOLECULAR BIOLOGY

For further volumes:
http://www.springer.com/series/7651

For over 35 years, biological scientists have come to rely on the research protocols and methodologies in the critically acclaimed *Methods in Molecular Biology* series. The series was the first to introduce the step-by-step protocols approach that has become the standard in all biomedical protocol publishing. Each protocol is provided in readily-reproducible step-by step fashion, opening with an introductory overview, a list of the materials and reagents needed to complete the experiment, and followed by a detailed procedure that is supported with a helpful notes section offering tips and tricks of the trade as well as troubleshooting advice. These hallmark features were introduced by series editor Dr. John Walker and constitute the key ingredient in each and every volume of the *Methods in Molecular Biology* series. Tested and trusted, comprehensive and reliable, all protocols from the series are indexed in PubMed.

Connexin Hemichannels

Methods and Protocols

Edited by

Fabio Mammano

Department of Physics and Astronomy "G. Galilei", University of Padova, Padova, Italy

Mauricio Retamal

Facultad de Medicina, Universidad del Desarrollo, Santiago, Chile

Humana Press

Editors
Fabio Mammano
Department of Physics and Astronomy
"G. Galilei"
University of Padova
Padova, Italy

Mauricio Retamal
Facultad de Medicina
Universidad del Desarrollo
Santiago, Chile

ISSN 1064-3745 ISSN 1940-6029 (electronic)
Methods in Molecular Biology
ISBN 978-1-0716-3841-5 ISBN 978-1-0716-3842-2 (eBook)
https://doi.org/10.1007/978-1-0716-3842-2

This Humana imprint is published by the registered company Springer Science+Business Media, LLC, part of Springer Nature.
The registered company address is: 1 New York Plaza, New York, NY 10004, U.S.A.

Paper in this product is recyclable.

Preface

Connexins are membrane proteins that form hexameric plasma membrane channels termed connexons or hemichannels (HCs) [1, 2]. When open, HCs mediate the diffusive release of paracrine messengers, most importantly ATP [3–9], but also glutamate and prostaglandins, NAD(+) or glutathione [10]. Solute permeation through HCs has been scrutinized in numerous studies by experimental methods (see, e.g., [11–16]) as well as molecular dynamics simulations [17–22]. Based on pore size estimates, the list of permeating molecules could potentially cover >35,000 members of the Human Metabolome Database[1] [23]. In several instances, molecules released through open HCs contribute to paracrine and autocrine signalling pathways during inflammation and cell death [24–27].

At least 29 dominant mutations in 8 human Cx genes (*GJA1, GJA3, GJA5, GJA5, GJA8, GJB1, GJB2, GJB3, GJB6*) generate HCs with augmented activity that have been implicated in 11 monogenic diseases, mostly rare, affecting facial appearance, dentition (small and carious teeth), eyes (microphthalmia, microcornea), fingers (syndactyly), inner ear (deafness), skin (various disorders), lens (cataracts), nervous system (myelination defects), and heart (atrial fibrillation). In several instances, aberrant connexin HCs cause an unbalance of Ca^{2+} signalling and/or augmented release of ATP which, acting as deregulated paracrine factor, compromises tissue homeostasis [28–31]. A limited number of mouse models, which carry some of these mutations, have been generated by different laboratories using different targeting strategies and genetic backgrounds.

In addition, numerous currently incurable diseases have been linked to the opening of wild-type HCs, including atherosclerosis, diabetes, ischemia/stroke, myocardial ischemia/reperfusion, cardiomyopathy, chronic kidney disease, microbial gut infection, HIV infection of astrocytes, acute liver failure, liver fibrosis and cirrhosis, nonalcoholic steatohepatitis, Alzheimer's disease, and epilepsy. Thus, HCs are important drug targets and their study is essential to identify new compounds capable of modulating HC function selectively and specifically.

Padova, Italy ***Fabio Mammano***
Santiago, Chile ***Mauricio Retamal***

References

1. DeVries SH, Schwartz EA (1992) Hemi-gap-junction channels in solitary horizontal cells of the catfish retina. J Physiol 445: 201–30
2. Paul DL, Ebihara L, Takemoto LJ, Swenson KI, Goodenough DA (1991) Connexin46, a novel lens gap junction protein, induces voltage-gated currents in nonjunctional plasma membrane of Xenopus oocytes. J Cell Biol 115(4): 1077–89
3. Toma I, Bansal E, Meer EJ, Kang JJ, Vargas SL, Peti-Peterdi J(2008) Connexin 40 and ATP-dependent intercellular calcium wave in renal glomerular endothelial cells. Am J Physiol Regul Integr Comp Physiol 294(6):R1769–76

[1] http://www.hmdb.ca/structures/search/metabolites/mass

4. Cotrina ML, Lin JH, Alves-Rodrigues A, Liu S, Li J, Azmi-Ghadimi H, Kang J, Naus CC, Nedergaard M (1998) Connexins regulate calcium signaling by controlling ATP release. Proc Natl Acad Sci U S A 95(26): 15735–40
5. Stout CE, Costantin JL, Naus CC, Charles AC (2002) Intercellular calcium signaling in astrocytes via ATP release through connexin hemichannels. J Biol Chem 277(12): 10482–8
6. Pearson RA, Dale N, Llaudet E, Mobbs P (2005) ATP released via gap junction hemichannels from the pigment epithelium regulates neural retinal progenitor proliferation. Neuron 46(5):731–44
7. Ho CL, Yang CY, Lin WJ, Lin CH (2013) Ecto-nucleoside triphosphate diphosphohydrolase 2 modulates local ATP-induced calcium signaling in human HaCaT keratinocytes. PLoS ONE 8(3):e57666
8. Anselmi F, Hernandez VH, Crispino G, Seydel A, Ortolano S, Roper SD, Kessaris N, Richardson W, Rickheit G, Filippov MA, Monyer H, Mammano F (2008) ATP release through connexin hemichannels and gap junction transfer of second messengers propagate Ca2+ signals across the inner ear. Proc Natl Acad Sci U S A 105(48):18770–5
9. Mazzarda F, D'Elia A, Massari R, De Ninno A, Bertani FR, Businaro L, Ziraldo G, Zorzi V, Nardin C, Peres C, Chiani F, Tettey-Matey A, Raspa M, Scavizzi F, Soluri A, Salvatore AM, Yang J, Mammano F (2016) Organ-on-chip model shows that ATP release through connexin hemichannels drives spontaneous Ca(2+) signaling in non-sensory cells of the greater epithelial ridge in the developing cochlea. Lab Chip 20(16):3011–3023
10. Wang N, De Bock M, Decrock E, Bol M, Gadicherla A, Vinken M, Rogiers V, Bukauskas FF, Bultynck G, Leybaert L (2013) Paracrine signaling through plasma membrane hemichannels. Bioch et Biophys Acta 1828(1):35–50
11. Trexler EB, Bukauskas FF, Kronengold J, Bargiello TA, Verselis VK (2000) The first extracellular loop domain is a major determinant of charge selectivity in connexin46 channels. Biophys J 79(6):3036–51
12. Orellana JA, Diaz E, Schalper KA, Vargas AA, Bennett MV, Saez JC (2011) Cation permeation through connexin 43 hemichannels is cooperative, competitive and saturable with parameters depending on the permeant species. Biochem Biophys Res Commun 409(4):603–9
13. Saez JC, Leybaert L (2014) Hunting for connexin hemichannels. FEBS Lett 588(8):1205–11
14. Nielsen BS, Alstrom JS, Nicholson BJ, Nielsen MS, MacAulay N (2017) Permeant-specific gating of connexin 30 hemichannels. J Biol Chem 292(49):19999–20009
15. Saez JC, Vargas AA, Hernandez DE, Ortiz FC, Giaume C, Orellana JA (2020) Permeation of molecules through astroglial connexin 43 hemichannels is modulated by cytokines with parameters depending on the permeant species. Int J Mol Sci 21(11)
16. Ahmad S, Diez JA, George CH, Evans WH (1999) Synthesis and assembly of connexins in vitro into homomeric and heteromeric functional gap junction hemichannels. Biochem J 339(Pt 2):247–53
17. Zonta F, Polles G, Zanotti G, Mammano F (2012) Permeation pathway of homomeric connexin 26 and connexin 30 channels investigated by molecular dynamics. J Biomol Struct Dyn 29(5):985–98
18. Zonta F, Polles G, Sanasi MF, Bortolozzi M, Mammano F (2013) The 3.5 angstrom X-ray structure of the human connexin26 gap junction channel is unlikely that of a fully open channel. Cell Commun Signal 11(1):15
19. Zonta F, Buratto D, Cassini C, Bortolozzi M, Mammano F (2014) Molecular dynamics simulations highlight structural and functional alterations in deafness-related M34T mutation of connexin 26. Front Physiol 5:85
20. Zonta F, Buratto D, Crispino G, Carrer A, Bruno F, Yang G, Mammano F, Pantano S (2018) Cues to opening mechanisms from in silico electric field excitation of Cx26 hemichannel and in vitro mutagenesis studies in HeLa transfectans. Front Mol Neurosci 11:170
21. Kwon T, Harris AL, Rossi A, Bargiello TA (2011) Molecular dynamics simulations of the Cx26 hemichannel: evaluation of structural models with Brownian dynamics. J Gen Physiol 138(5):475–93
22. Nielsen BS, Zonta F, Farkas T, Litman T, Nielsen MS, MacAulay N (2019) Structural determinants underlying permeant discrimination of the Cx43 hemichannel. J Biol Chem 294(45):16789–16803
23. Esseltine JL, Laird DW (2016) Next-generation connexin and pannexin cell biology. Trends cell Biol 26(12):944–955
24. Decrock E, Vinken M, De Vuyst E, Krysko DV, D'Herde K, Vanhaecke T, Vandenabeele P, Rogiers V, Leybaert L (2009) Connexin-related signaling in cell death: to live or let die? Cell Death Differ 16(4): 524–36

25. Chandrasekhar A, Bera AK (2012) Hemichannels: permeants and their effect on development, physiology and death. Cell Biochem Funct 30(2):89–100
26. Vinken M (2015) Connexin hemichannels: novel mediators of toxicity. Arch Toxicol 89(1):143–5
27. Dosch M, Gerber J, Jebbawi F, Beldi G (2018) Mechanisms of ATP release by inflammatory cells. Int J Mol Sci 19(4)
28. Retamal MA, Reyes EP, Garcia IE, Pinto B, Martinez AD, Gonzalez C (2015) Diseases associated with leaky hemichannels. Front Cell Neurosci 9:267
29. Garcia IE, Bosen F, Mujica P, Pupo A, Flores-Munoz C, Jara O, Gonzalez C, Willecke K, Martinez AD (2016) From hyperactive Connexin26 hemichannels to impairments in epidermal calcium gradient and permeability barrier in the keratitis-ichthyosis-deafness syndrome. J Invest Dermatol 136(3):574–83
30. Leybaert L, Lampe PD, Dhein S, Kwak BR, Ferdinandy P, Beyer EC, Laird DW, Naus CC, Green CR, Schulz R (2017) Connexins in cardiovascular and neurovascular health and disease: pharmacological implications. Pharmacol Rev 69(4):396–478
31. Srinivas M, Verselis VK, White TW (2018) Human diseases associated with connexin mutations. Biochim et Biophys Acta 1860(1):192–201

Contents

Contributors

FRANCISCA M. ACOSTA • *Department of Biochemistry and Structural Biology, University of Texas Health Science Center at San Antonio, San Antonio, TX, USA*

GUILLERMO A. ALTENBERG • *Department of Cell Physiology and Molecular Biophysics, and Center for Membrane Protein Research, Texas Tech University Health Sciences Center, Lubbock, TX, USA*

ERVA BAYRAKTAR • *Department of Physics and Astronomy "G. Galilei", University of Padua, Padova, Italy; Veneto Institute of Molecular Medicine (VIMM), Padova, Italy*

MARIO BORTOLOZZI • *Department of Physics and Astronomy "G. Galilei", University of Padua, Padova, Italy; Veneto Institute of Molecular Medicine (VIMM), Padova, Italy*

KARINE BRECKPOT • *Laboratory for Molecular and Cellular Therapy, Department of Biomedical Sciences, Vrije Universiteit Brussel, Brussels, Belgium*

AXELLE COOREMAN • *Entity of In Vitro Toxicology and Dermato-Cosmetology, Department of Pharmaceutical and Pharmacological Sciences, Vrije Universiteit Brussel, Brussels, Belgium*

YANNICK DE VLAEMINCK • *Laboratory for Molecular and Cellular Therapy, Department of Biomedical Sciences, Vrije Universiteit Brussel, Brussels, Belgium*

CHIARA DI PIETRO • *Institute of Biochemistry and Cell Biology, Italian National Research Council, Rome, Italy*

VIOLA DONATI • *Institute of Biochemistry and Cell Biology, Italian National Research Council, Rome, Italy; Department of Biomedical Sciences, University of Padova, Padova, Italy*

AINOA FERNANDEZ-OLIVARES • *Programa de Comunicación Celular en Cáncer, Facultad de Medicina Clínica Alemana, Universidad del Desarrollo, Santiago, Chile*

MARIANA C. FIORI • *Clinical Research Institute, Texas Tech University Health Sciences Center, Lubbock, TX, USA*

ANÍBAL GARCÍA • *Instituto de Neurociencias, Centro Interdisciplinario de Neurociencias de Valparaíso (CINV), Universidad de Valparaíso, Valparaíso, Chile*

HENRIQUE GIRÃO • *Univ Coimbra, Coimbra Institute for Clinical and Biomedical Research (iCBR), Faculty of Medicine, Coimbra, Portugal; Univ Coimbra, Center for Innovative Biomedicine and Biotechnology (CIBB), Coimbra, Portugal; Clinical Academic Centre of Coimbra (CACC), Coimbra, Portugal*

JORGE GONZÁLEZ • *Molecular Parasitology Unit, Faculty of Health Sciences, Medical Technology Department, Universidad de Antofagasta, Antofagasta, Chile*

SUMIN GU • *Department of Biochemistry and Structural Biology, University of Texas Health Science Center at San Antonio, San Antonio, TX, USA*

JEAN X. JIANG • *Department of Biochemistry and Structural Biology, University of Texas Health Science Center at San Antonio, San Antonio, TX, USA*

PRASHANT KADAM • *Entity of In Vitro Toxicology and Dermato-Cosmetology, Department of Pharmaceutical and Pharmacological Sciences, Vrije Universiteit Brussel, Brussels, Belgium*

VOLODYMYR M. KORKHOV • *Laboratory of Biomolecular Research, Paul Scherrer Institute, Villigen, Switzerland; Institute of Molecular Biology and Biophysics, ETH, Zurich, Switzerland*

DALE W. LAIRD • *Department of Anatomy and Cell Biology, University of Western Ontario, London, ON, Canada; Department of Physiology and Pharmacology, University of Western Ontario, London, ON, Canada*
PIA LAVRIHA • *Laboratory of Biomolecular Research, Paul Scherrer Institute, Villigen, Switzerland*
STEPHANIE E. LEIGHTON • *Department of Anatomy and Cell Biology, University of Western Ontario, London, ON, Canada*
SERGIU A. LUCACIU • *Department of Anatomy and Cell Biology, University of Western Ontario, London, ON, Canada; Department of Physiology and Pharmacology, University of Western Ontario, London, ON, Canada*
FABIO MAMMANO • *Institute of Biochemistry and Cell Biology, Italian National Research Council, Rome, Italy; Department of Physics and Astronomy "G. Galilei", University of Padova, Padova, Italy*
DANIELA MARAZZITI • *Institute of Biochemistry and Cell Biology, Italian National Research Council, Rome, Italy*
TÂNIA MARTINS-MARQUES • *Univ Coimbra, Coimbra Institute for Clinical and Biomedical Research (iCBR), Faculty of Medicine, Coimbra, Portugal; Univ Coimbra, Center for Innovative Biomedicine and Biotechnology (CIBB), Coimbra, Portugal; Clinical Academic Centre of Coimbra (CACC), Coimbra, Portugal*
CHIARA NARDIN • *Institute of Biochemistry and Cell Biology, Italian National Research Council, Rome, Italy*
NICOLÁS PALACIOS-PRADO • *Laboratorio de Neurobiología, Departamento de Fisiología, Pontificia Universidad Católica de Chile, Santiago, Chile*
SERGIO PANTANO • *Institut Pasteur de Montevideo, Montevideo, Uruguay*
CHIARA PERES • *Institute of Biochemistry and Cell Biology, Italian National Research Council, Rome, Italy*
CHAO QI • *Institute of Molecular Biology and Biophysics, ETH, Zurich, Switzerland*
MAURICIO A. RETAMAL • *Programa de Comunicación Celular en Cáncer, Facultad de Medicina Clínica Alemana, Universidad del Desarrollo, Santiago, Chile*
MANUEL A. RIQUELME • *Department of Biochemistry and Structural Biology, University of Texas Health Science Center at San Antonio, San Antonio, TX, USA*
JUAN C. SÁEZ • *Instituto de Neurociencias, Centro Interdisciplinario de Neurociencias de Valparaíso (CINV), Universidad de Valparaíso, Valparaíso, Chile; Centro Interdisciplinario de Neurociencias de Valparaíso, Instituto de Neurociencias, Universidad de Valparaíso, Valparaíso, Chile*
JULEN SANZ-SERRANO • *Entity of In Vitro Toxicology and Dermato-Cosmetology, Department of Pharmaceutical and Pharmacological Sciences, Vrije Universiteit Brussel, Brussels, Belgium*
ANDRÉS TABERNILLA • *Entity of In Vitro Toxicology and Dermato-Cosmetology, Department of Pharmaceutical and Pharmacological Sciences, Vrije Universiteit Brussel, Brussels, Belgium*
ABRAHAM TETTEY-MATEY • *Institute of Biochemistry and Cell Biology, Italian National Research Council, Rome, Italy*
RAF VAN CAMPENHOUT • *Entity of In Vitro Toxicology and Dermato-Cosmetology, Department of Pharmaceutical and Pharmacological Sciences, Vrije Universiteit Brussel, Brussels, Belgium*
JOSÉ LUIS VEGA • *Facultad de Ciencias Biológicas, Departmento de Fisiología, Universidad de Concepción, Concepción, Chile*

TERESA VERGARA • *Laboratorio de Neurobiología, Departamento de Fisiología, Pontificia Universidad Católica de Chile, Santiago, Chile*
MATHIEU VINKEN • *Entity of In Vitro Toxicology and Dermato-Cosmetology, Department of Pharmaceutical and Pharmacological Sciences, Vrije Universiteit Brussel, Brussels, Belgium*
FRANCESCO ZONTA • *Department of Biological Sciences, School of Science, Xi'an Jiaotong-Liverpool University, Suzhou, China*

Chapter 1

Purification, Reconstitution, and Functional Analysis of Connexin Hemichannels

Mariana C. Fiori and Guillermo A. Altenberg

Abstract

Connexins are the proteins that form the gap junction channels that are essential for cell-to-cell communication. These channels are formed by head-to-head docking of hemichannels (each from one of two adjacent cells). Free "undocked" hemichannels at the plasma membrane are mostly closed, although they are still important under physiological conditions. However, abnormal and sustained increase in hemichannel activity due to connexin mutations or acquired conditions can produce or contribute to cell damage. For example, mutations of Cx26, a connexin isoform, can increase hemichannel activity and cause deafness. Studies using purified isolated systems under well-controlled conditions are essential for a full understanding of molecular mechanisms of hemichannel function under normal conditions and in disease, and here, we present methodology for the expression, purification, and functional analysis of hemichannels formed by Cx26.

Key words Calcium, Channel, Cx26, Cx43, Fluorescent probe, Gap junction channel, Membrane transport, Permeability

1 Introduction

Human connexins include 21 isoforms [1, 2]. These proteins can form gap junction channels and hemichannels (HCs), where HCs are connexin hexamers that can dock head-to-head to form a gap junction channel (Fig. 1) [1, 3, 4]. In the gap junction channels, each of two adjacent cells contribute one HC, and the resulting gap junction channels communicate the cytoplasms of the cells, which is essential for chemical and electric coupling, as HCs and gap junction channels are permeable to small hydrophilic molecules of up to ~1 kDa [5–7]. Under normal conditions, free "undocked" HCs at the plasma membrane participate in autocrine and paracrine signaling by mediating the efflux of signaling compounds such as ATP, NAD^+, and prostaglandins [8–11]. However, abnormal sustained opening of HCs can produce or contribute to the cell damage that occurs with connexin mutations or in conditions such as those

Fabio Mammano and Mauricio Retamal (eds.), *Connexin Hemichannels: Methods and Protocols*, Methods in Molecular Biology, vol. 2801, https://doi.org/10.1007/978-1-0716-3842-2_1,

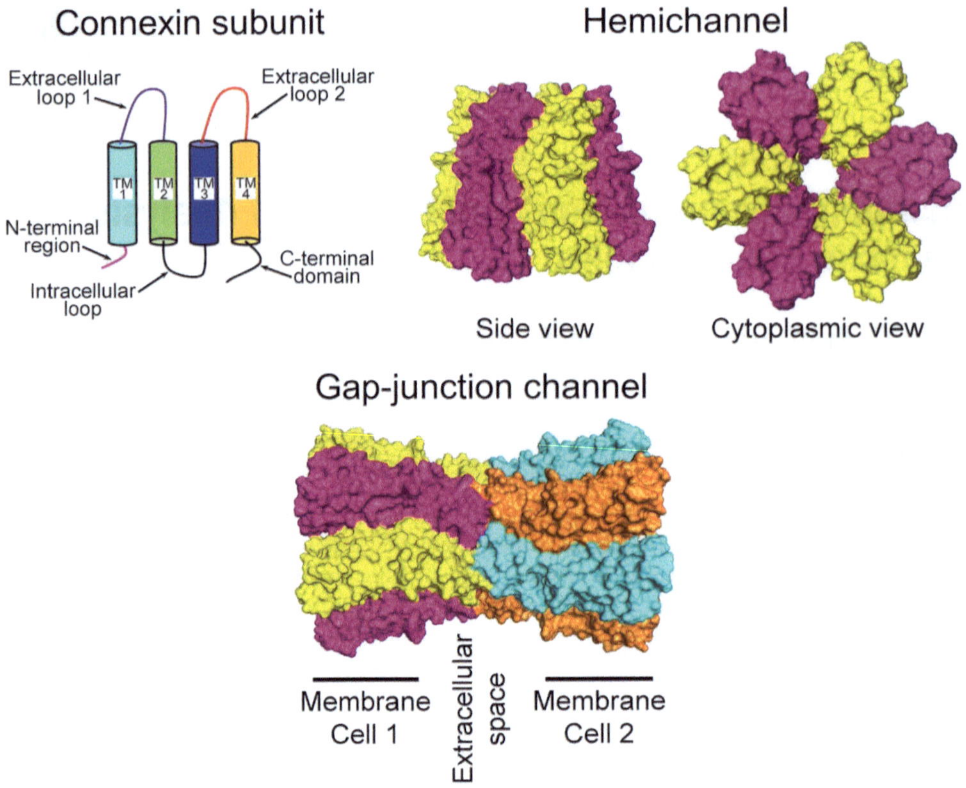

Fig. 1 Connexins, hemichannels, and gap-junctional channels. Top left: schematic representations of a connexin subunit (monomer). TM1–4: transmembrane helices 1 to 4. Top right: surface representation of a connexin hemichannel. Subunits are depicted in alternating colors. Bottom: side view surface representation of a gap-junctional channel. Subunits are depicted in alternating colors (magenta and yellow in one hemichannel, and cyan and orange in the other hemichannel). (Modified with permission from Natha et al. [47])

characterized by inflammation, hypoxia, and oxidative stress, including cardiac infarcts and stroke (Fig. 2) [9, 12–29].

Studies of HCs not only have physiological and pathophysiological importance in their own right, but also contribute to our understanding of gap junction channels. This is because gap junction channels formed by connexins such as Cx26 behave as two HCs in series [22, 30]. Even though Cx43 HCs are activated only at positive intracellular voltages, these HCs display single-channel conductance and kinetics consistent with that of Cx43 gap junction channels, and several mechanisms of gap junction channel and HC regulation are similar [14, 31–39]. Therefore, understanding the mechanisms of regulation of HCs also contributes to our knowledge of gap junction channel regulation.

The overwhelming majority of research on HCs has been done on complex systems such as cells, organs, and organisms. Experiments in these systems are critical to understand processes in health

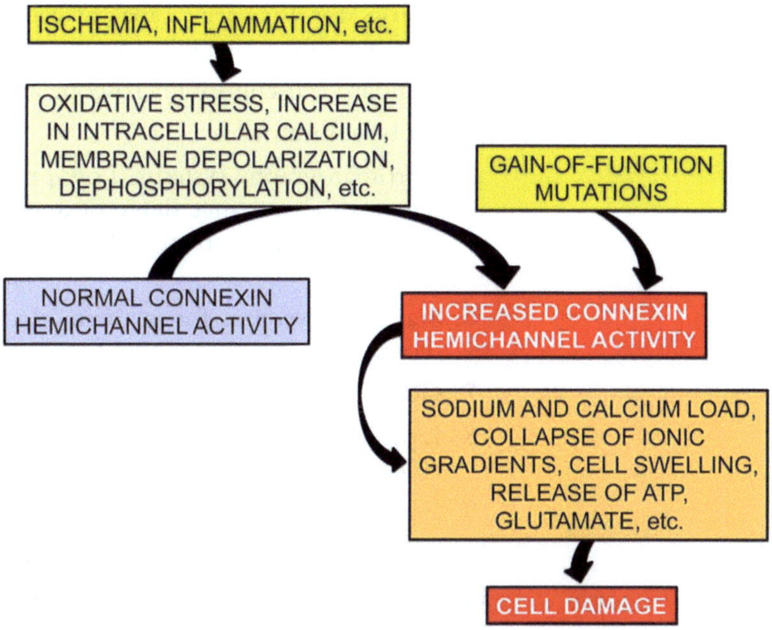

Fig. 2 Simplified scheme of the cell damage by increased hemichannel activity in genetic and acquired disorders. (Modified with permission from Natha et al. [47])

and disease, but their complexity often hinders the discovery of details at the molecular level. Thus, studies in purified isolated systems under well-controlled conditions are essential for a full understanding of molecular mechanisms under normal conditions and in disease.

Here, we present methodology to assess purified HCs with a focus on the expression, purification, reconstitution, and functional assays of HCs formed by Cx26, a connexin isoform of medical importance, as its mutations are the most common cause of genetic deafness.

2 Materials

Here, we list details on the equipment, materials, and chemicals, but only give specific brand information for those that are not common or we consider important.

2.1 Equipment

1. Incubators. Standard incubators set at 37 °C for bacterial agar plates and at 26 °C for insect cell culture flasks, and a shaker incubator set at 26 °C for growing insect cells in suspension.
2. Centrifuges and rotors. Refrigerated centrifuges, including, at a minimum, a table-top microcentrifuge, a general-purpose centrifuge, a large capacity centrifuge, and an ultracentrifuge.

Rotors for the large capacity centrifuge and rotors for the ultracentrifuge, including a swinging bucket rotor.

3. Standard cell-culture microscope and cell counter.
4. Probe sonicator.
5. Chromatography systems. Low-pressure chromatography system with absorbance detector and medium-pressure chromatography system with absorbance and fluorescence detectors in series.
6. Standard systems for protein electrophoresis and Western blots.
7. Gel imager capable of infrared imaging.
8. Lipid drying system. Argon gas directed to blow on lipids in chloroform inside a glass tube placed in a chemical hood.
9. Liposome extruder. Avanti Polar Lipids Mini-Extruder.
10. Spectrophotometer, spectrofluorometer, and luminometer.
11. Stopped-flow system. Applied Photophysics SX20 Stopped-Flow Spectrometer.
12. Osmometer.
13. Plate reader.
14. Dynamic light scattering instrument for molecular sizing.

2.2 Solutions

All solutions prepared at room temperature with reagent-grade chemicals and distilled water treated on an ion-exchange water purification system.

1. Pellet resuspension solution: 1 mM sodium bicarbonate solution containing 1 mM phenylmethylsulfonyl fluoride (PMSF).
2. Solubilization solution: 2.5% *n*-dodecyl-*β*-D-maltoside (DDM) in 2 M NaCl, 2 mM ethylene glycol-bis(2-aminoethylether)-*N*,*N*,*N*′,*N*′-tetraacetic acid (EGTA), 2 mM dithiothreitol (DTT), 10% glycerol, 1 mM PMSF, and 10 mM glycine/NaOH, pH 8.
3. Dilution buffer: 5 mM imidazole, 10% glycerol, 1 mM PMSF, 0.03% DDM, and 50 mM HEPES/NaOH, pH 8.0.
4. Washing buffer: 10 mM KCl, 10% glycerol, 20 mM imidazole, 0.03% DDM, and 10 mM HEPES/KOH, pH 7.4.
5. Elution buffer: 10 mM KCl, 10% glycerol, 250 mM imidazole, 0.03% DDM, and 10 mM HEPES/KOH, pH 7.4.
6. Gel filtration buffer: 0.03% DDM, 150 mM NaCl, 10% glycerol, and 10 mM HEPES/NaOH, pH 7.4.
7. Reconstitution buffer: phosphatydilcholine (PC) and phosphatydilserine (PS) at a 2:1 ratio (w/w) in 100 mM KCl, 0.1 mM EGTA, 2.3% *n*-octyl-*β*-D-glucopyranoside (OG), and 25 mM HEPES, pH 7.6, at 75 mL of buffer/mg of lipid.

8. Dialysis buffer: 100 mM KCl, 0.1 mM EGTA, and 25 mM HEPES, pH 7.6, with 5 mL of a 50% (w/v) suspension of Bio-Rad Bio-Beads SM-2 *per* liter of solution.
9. Reconstitution buffer for sucrose transport experiments: phosphatidylethanolamine labeled with lissamine rhodamine B (PE-R) combined with PC and PS at a PC:PS:PE-R ratio of 2:1:0.05 (w/w/w) in 10 mM KCl, 0.1 mM EGTA, 459 mM urea, and 10 mM HEPES, pH 7.6.
10. Transport buffer: 100 mM KCl, 0.1 mM EGTA, 25 mM HEPES/KOH, pH 7.6.
11. Low pH transport buffer: transport buffer titrated with HCl to pH 6.4.
12. Na^+-free buffer: 70 mM KCl, 0.1 mM EGTA, 30 mM *N*-methyl-D-glucamonium (NMDG), and 25 mM HEPES/KOH, pH 7.6.
13. 15 mM Na^+ buffer: 70 mM KCl, 0.1 mM EGTA, 15 mM NaCl, 15 mM NMDG, and 25 mM HEPES/KOH, pH 7.6.
14. Luciferin/luciferase reaction solution: 18 nM luciferase (QuantiLum *Photinus pyralis* recombinant luciferase from Promega), 610 mM luciferin (GoldBio), 0.5 mM ethylenediaminetetraacetic acid (EDTA), and 5 mM $MgSO_4$.

2.3 Commercial Reagents

1. Insect cells and cell culture media. SF9 insect cells; HyClone CCM3 medium; Sf-900 III SFM complete growth medium; Grace's Insect Cell Culture Medium, unsupplemented.
2. Bac-to-Bac Baculovirus Expression System (Thermo Fisher Scientific).
3. Antibiotics and isopropyl β-D-1-thiogalactopyranoside (IPTG).
4. LB agar plates containing 50 μg/mL kanamycin, 7 μg/mL gentamicin, 10 μg/mL tetracycline, 100 μg/mL Bluo-gal, and 40 μg/mL IPTG.
5. S.O.C. medium.
6. DNA isolation kit.
7. Resins for immobilized metal affinity chromatography (IMAC) (e.g., Talon Superflow Co^{2+} resin from Takara Bio USA; Ni^{2+}-NTA resin from Qiagen, Valencia, CA).
8. Detergents. DDM and OG.
9. Lipids. All lipids purchased as chloroform stocks. These are PC and PS and fluorescein-labeled phospholipid N-(fluorescein-5-thiocarbamoyl)-1,2-dihexadecanoyl-sn-glycero-3-phosphoethanolamine (DHPE-Fl; head group-labeled; Thermo Fisher Scientific) and PE-R (Thermo Fisher Scientific).

10. Bio-Beads SM-2 (Bio-Rad).
11. Centrifugal concentrators (100 kDa cutoff).
12. Size-exclusion chromatography columns.
13. Desalting/buffer exchange columns.
14. Dialysis membrane (e.g., 6000–8000 molecular mass cutoff).
15. Antibodies for Western blots. Primary antibodies directed against the intracellular loop and N-terminal region of Cx26 (Zymed Laboratories) and the His tag (anti-Penta-His; Qiagen). Secondary antibodies: goat anti-rabbit IRDye 800 (Li-Cor Biosciences) for the anti-Cx26 antibodies and goat anti-mouse Alexa Fluor 680 (Thermo Fisher) for the anti-His antibody.
16. Restriction grade thrombin.
17. Fluorescent probes. Alexa Fluor 350 (AF350), Alexa Fluor 647 (AF647), Ca^{2+}-sensitive probe Fluo-5 N (pentapotassium salt), and Na^{+}-sensitive probe 1,3-benzenedicarboxylic acid, 4,4′-[1,4,10-trioxa-7,13-diazacyclopentadecane-7,13-diylbis(5-methoxy-6,2-benzofurandiyl)]bis (SBFI, tetraammonium salt).

3 Methods

The methods are divided in several sections: Generation of recombinant baculoviruses, expression of connexins in insect cells, purification of connexin hemichannels, reconstitution of connexin hemichannels, and functional assays of reconstituted connexin hemichannels. Publications from our laboratory that include the methodology presented here can be consulted for details and applications [34, 40–42].

3.1 Generation of Recombinant Baculoviruses

1. Engineer wild-type human Cx26 followed by a poly-His tag (six His) at the C-terminal end. One possibility is to amplify human Cx26 by PCR with the addition of the poly-His tag and *EcoRI* and *XbaI* flanking sequences for cloning into the baculovirus transfer vector pFastBac 1 [41].
2. Transform *E coli* DH10Bac with the purified pFastBac-Cx26 construct following the Bac-to-Bac Baculovirus Expression system's instructions for transposition into the bacmid to generate the baculovirus [43, 44]. Plate the transformation mix into LB agar plates containing 50 μg/mL kanamycin, 7 μg/mL gentamicin, 10 μg/mL tetracycline, 100 μg/mL Bluo-gal, and 40 μg/mL IPTG, and select white colonies containing the recombinant bacmid (*see* **Note 1**).

3. Inoculate the colonies with the white phenotype (*see* **Note 2**) into a 25 mL liquid LB culture containing 50 μg/mL kanamycin, 7 μg/mL gentamicin, 10 μg/mL tetracycline, and 40 μg/mL IPTG, and growth them overnight.
4. Isolate the bacmid DNA using a DNA isolation kit (*see* **Note 3**).
5. Verify the presence of the gene of interest transposed into the bacmid by PCR using M13 primers.
6. Transfect Sf9 insect cells with the bacmid. For this, plate the cells in log phase (~2×10^6 cells/mL) at a density of 8×10^5 cells/well in 6-well tissue-culture plates. Incubate them at 26 °C in Grace's medium supplemented with 10% fetal calf serum and 0.05 mg/mL gentamycin, essentially as described [34, 41]. Allow the cells to attach for 15 min at room temperature in a tissue culture hood, and then replace the medium with a medium mix containing Cellfectin II and the bacmid DNA, carefully added dropwise onto the cells.
7. After incubation of the cells at 26 °C for 3–5 h, replace the transfection mixture with 2 mL of Sf-900 III SFM complete growth medium, and incubate the plates at 26 °C until signs of viral infection appear (*see* **Note 4**).
8. Once the cells appear infected (i.e., showed characteristics typical of late to very late infection), collect the medium containing virus from each well (~2 mL/well) in 15 mL snap-cap tubes, and centrifuge at 500 × *g* for 5 min to remove cells and large debris. Transfer the clarified supernatant to fresh sterile tubes. This is the P1 viral stock.
9. Amplify the P1 viral stock into P2 and then into P3 stocks to obtain the highest viral titer and enough volume of the P3 virus to express the protein at large scale (*see* **Note 5**).
10. Analyze Cx26 expression and optimize conditions if needed. For this, grow Sf9 cells at 26 °C in suspension, shaken at 125 rpm in 250 mL baffled flasks containing 50 mL of HyClone CCM3 medium. Infect the cells at a density of 2×10^6 cells/mL using a multiplicity of infection (MOI) of 2. Collect 1 mL samples at specified time intervals to measure viability by Trypan blue staining.
11. Determine Cx26 expression by Western blotting at the selected times and choose the optimal expression conditions from the cell viability vs. Cx26 expression.

3.2 Expression of Connexins in Insect Cells

1. Express Sf9 cells in suspension, in 2-L baffled flasks containing 750 mL of HyClone CCM3 medium [41]. Grow the cells at 26 °C, shaken at 125 rpm, and infect them at a concentration of 2×10^6 cells/mL, using an MOI of 2.

2. Harvest the cells by centrifugation at 1000 × *g* for 10 min at 4 °C when cell viability, determined by Trypan blue staining, is ~40%, which generally occurs ~48 h postinfection.
3. The cell pellet is ready for use fresh to purify the protein or for storage at −80 °C.

3.3 Purification of Connexin Hemichannels

1. You can start the purification by resuspending fresh pellets or defrosted pellets stored at −80 °C in pellet resuspension solution, essentially as published [34, 40, 41].
2. Homogenize the cells with a Dounce glass homogenizer and centrifuge the suspension at 1000 × *g* for 5 min at 4 °C.
3. Alkali-extract the pellet by addition of NaOH to 60 mM, followed by incubation on ice for 30 min.
4. Centrifuge the alkali-extracted samples at 26,000 × *g* for 30 min and resuspend the resulting membrane pellet in solubilization solution at a protein concentration of ~2 mg/mL.
5. Sonicate the suspension with a tip probe sonicator and then incubate the sample for 2 h with gentle rotation at 4 °C.
6. Ultracentrifuge the sample at 100,000 × *g* for 30 min, and then add 2.5 volumes of dilution buffer to the solubilized material (supernatant). Load this diluted sample at a rate of 0.5 mL/min onto a Talon Co^{2+} column pre-equilibrated with dilution buffer.
7. Wash the column with 10 column volumes of washing buffer and proceed with the elution by switching to elution buffer.
8. Subject the IMAC-purified Cx26 sample to size-exclusion chromatography. For these experiments, inject Cx26 (500 μL at ~0.5 mg/mL) in gel filtration buffer into the column equilibrated with the same buffer, and run at a flow rate of 0.5 mL/min. You can calculate the apparent molecular mass from the linear relationship between the partition coefficient K_{av} and the log of the molecular mass, using the following standards, run under the same conditions: 440 kDa (ferritin), 158 kDa (aldolase), 43 kDa (ovalbumin), and 13.7 kDa (ribonuclease A); blue dextran 2000 (~2000 kDa) can be used to determine the void volume.
9. Determine protein concentration and Cx26 presence by Western blotting.
10. You can analyze the highly purified Cx26 (peak collected from the gel filtration run) by dynamic light scattering to confirm apparent size and monodispersity.
11. Store the purified protein in aliquots at −80 °C.

3.4 Reconstitution of Connexin Hemichannels

1. Perform the reconstitution in a mixture of PC and PS at a 2:1 ratio (w/w), as described below [34, 40, 41].
2. Lyophilize the lipids from chloroform stocks overnight, under argon, mixed at the indicated ratio.
3. Rehydrate the dry lipids film in reconstitution buffer at 75 μL of buffer/mg of lipid. Warm the suspension to 37 °C until it becomes transparent (*see* **Note 6**).
4. Add purified Cx26 to the lipid–detergent mixture and dialyze the sample for 36 h at room temperature against 2 L of detergent-free buffer containing 10 mL of a 50% (w/v) suspension of Bio-Beads SM-2. Change the dialysis buffer twice during the 36-h period, after approximately 5 and 24 h. Alternatively, you can produce liposomes and proteoliposomes by gel filtration [42].
5. Produce unilamellar liposomes and proteoliposomes by extrusion, as described in **steps 6–9**.
6. Load the dialysate sample into one of the Mini-Extruder gas--tight syringes and carefully place it into one end of the Mini-Extruder, opposite to another identical syringe from which it is separated by a filter holder assembly containing a 100 μm polycarbonate membrane (*see* **Note** 7).
7. Push the plunger of the filled syringe gently until the lipid solution passes through the filter and is completely transferred to the other syringe.
8. Transfer the solution back to the original syringe and repeat the two-step process 10 times (*see* **Note 8**).
9. If needed, estimate the proportion of reconstituted HCs with the extracellular domains facing out from thrombin digestion experiments. For these studies, you can use an engineered human Cx26 with a thrombin cleavage site between the end of the Cx26 sequence and the poly-His tag (Cx26T). You can introduce the thrombin site by exchange of the 742-bp *BamHI/XbaI* fragment of Cx26 with an identical PCR fragment, except for the addition of the sequence coding for the thrombin cleavage site (LVPRGS) [41]. Incubate Cx26T reconstituted in liposomes (typically 20 μg of protein) for 16 h at room temperature with 0.1 units of restriction grade thrombin and stop the reaction by adding SDS gel sample buffer. Estimate the percentage of HCs with the extracellular domains facing out from comparative densitometry analysis of the signal from immunoblots probed with an anti-His tag antibody (decrease in signal compared to the untreated sample (*see* **Note 9**)).

3.5 Functional Assays of Reconstituted Connexins Hemichannels

1. The transport-specific fractionation technique can be used for sucrose transport assays [34, 41, 45, 46]. For this, reconstitute Cx26 in liposomes containing traces of PE-R at a PC:PS:PE-R ratio of 2:1:0.05 (w/w/w) using reconstitution buffer for sucrose transport experiments.
   (a) Layer the liposomes or proteoliposomes containing PE-R onto a linear iso-osmolar sucrose gradient (0–400 mM sucrose with a reverse urea gradient) (Fig. 3a).

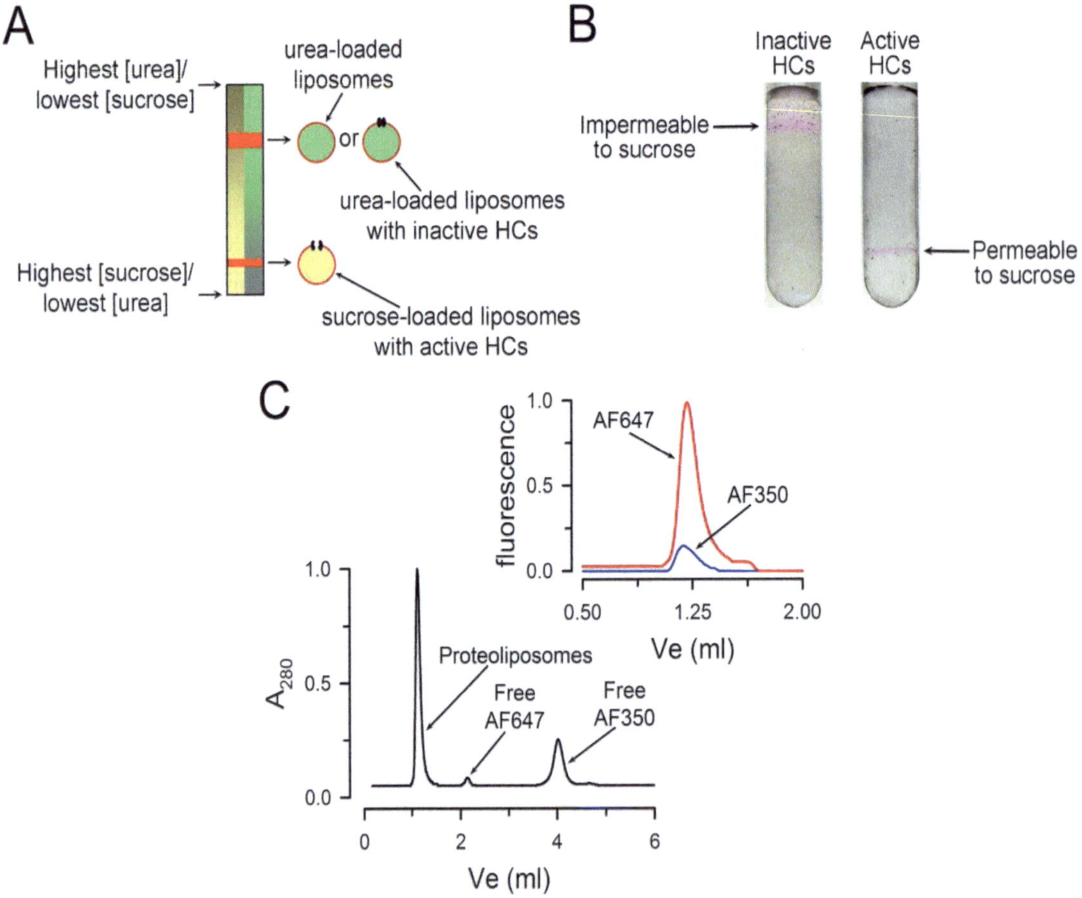

Fig. 3 Qualitative transport assays. (**a**) Schematic representation of the transport-specific assay. (**b**) Typical migration of liposomes containing purified HCs on an iso-osmolar linear sucrose density gradient. The images show the position of the rhodamine B-labeled proteoliposomes after centrifugation for proteoliposomes containing inactive (sucrose impermeable) or active sucrose permeable) HCs. (**c**) Permeability of reconstituted HCs to Alexa Fluor probes. Liposomes preloaded with Alexa Fluor 350 (AF350) or 647 (AF647) were run on the size-exclusion column to separate free extraliposomal probes from the probes retained inside the liposomes. A_{280} is the absorbance at 280 nm. Top right: AF350 and AF647 fluorescence associated with the liposomes containing Cx26. The data were normalized to the peak emission of parallel experiments with liposomes without Cx26. Panels **A**, **B**, and **C** were modified with permission from Fiori et al. reference [15], Bao et al. [34], and Fiori et al. [41], respectively

(b) Centrifuge the gradient on a swinging bucket rotor at 125,000 × *g* for 5 h at 4 °C.

(c) Determine the location of the liposomes from direct visualization and from the fluorescence of fractions collected from the top to the bottom of the gradient. For this, measure the rhodamine B fluorescence at 590 nm with excitation at 560 nm (Fig. 3a, b).

2. To assay the transport of fluorescent probes, load the liposomes and proteoliposomes with the fluorescent probes AF350 (250 mM; permeable through open Cx26 HCs) and/or AF647 (125 mM; impermeable through Cx26 HCs) in 100 mM KCl, 0.1 mM EGTA, and 25 mM HEPES/KOH, pH 7.6, and determine the fraction of the probes retained after size-exclusion chromatography [41, 42]. AF350 is permeable through Cx26 HCs and AF647 is not (molecular masses of ~350 and 1,300 Da, respectively).

(a) Load the probes into the liposomes and proteoliposomes by four freeze-thaw cycles. For each cycle, place the samples 1 min in liquid N_2 followed by 5 min in a bath at 27 °C (*see* **Note 10**).

(b) Remove the extraliposomal fluorescent probes by gel filtration on PD-10 columns, and then run the samples at 21–22 °C on a size-exclusion chromatography column using a chromatography system equipped with absorbance and multiwavelength fluorescence detectors.

(c) Determine the position of the liposomes by the absorbance at 280 nm, and the retention of dye trapped in liposomes by fluorescence, using excitation and emission wavelengths of 345 and 445 nm for AF350, and 650 and 670 nm for AF647, respectively (Fig. 3c).

3. For analysis of ATP transport, load the liposomes and proteoliposomes with 1 mM ATP, as described above, and assay the ATP retained by luminometry [41, 42].

(a) Remove the extraliposomal ATP by size-exclusion chromatography on PD-10 columns, combine 250 mL of the samples with 5 mL of the luciferin/luciferase mix, measure the luminescence of the sample, and compare it to ATP standards.

(b) Determine the ATP retained in the liposomes as the difference in luminescence between total ATP (after addition of Triton X-100 to 0.4%) and extraliposomal ATP (accessible to luciferase in the absence of Triton X-100).

4. For the Ca^{2+} transport experiments, load the HC impermeable low-affinity Ca^{2+}-sensitive probe Fluo-5 N (25 mM; impermeable through Cx26 HCs) into the liposomes, as described for the fluorescent probes above, and determine the increase in free [Ca^{2+}] upon increasing extraliposomal concentration from

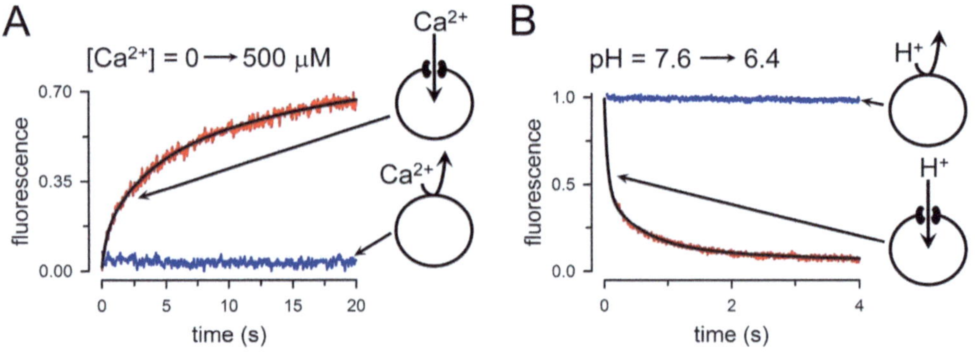

Fig. 4 Quantitative transport assays. (**a**) Rate of Ca^{2+} influx into liposomes containing purified Cx26 HCs. Extraliposomal free-[Ca^{2+}] was rapidly increased from a few nM to 500 μM by mixing in a stopped-flow cell, and the rate of increase in emission from Fluo-5 N trapped into the liposomes was followed. Red record: liposomes containing Cx26 HCs; blue record: liposomes without HCs. The black line corresponds to a multiexponential fit to the data that was used to calculate the rate of Ca^{2+} influx. (**b**) Rate of change in pH inside liposomes containing purified Cx26 HCs due to the influx of H^+ equivalents. Extraliposomal pH was rapidly reduced from 7.6 to 6.4 by mixing in a stopped-flow cell, and the rate of decrease in pH inside the liposomes was followed by the quenching of the fluorescein bound to the inner leaflet of the liposomes bilayer. Red record: liposomes containing Cx26 HCs; blue record: liposomes without HCs. The black line corresponds to a multiexponential fit to the data that was used to calculate the rate of liposome acidification. (Reproduced with permission from Fiori et al. [15]; modified with permission from Fiori et al. [41])

~10 nM to 500 μM [41, 42]. Perform this and the following experiments at 20 °C using a standard spectrofluorometer or a stopped-flow device. You can elevate free [Ca^{2+}] by manual addition of $CaCl_2$ to a cuvette in a fluorometer (excitation: 490 nm; emission: 520 nm), or by rapid mixing on a stopped-flow device, with excitation at 470 nm and emission collected through a 500-nm-long pass filter (Fig. 4a) [41, 42].

5. For Na^+ transport experiments, load the liposomes with the HC impermeable Na^+-sensitive probe SBFI (25 mM; impermeable through Cx26 HCs) in a buffer devoid of Na^+ (replaced with NMDG): 70 mM KCl, 0.1 mM EGTA, 30 mM NMDG chloride, and 25 mM HEPES/KOH, pH 7.6. Increase extraliposomal [Na^+] to 15 mM by replacing the NMDG chloride in the solution above with NaCl. For these experiments, use a stopped-flow spectrometer with excitation at 340 nm and emission measured through a 400-nm-long pass filter [41].

6. The estimate of transport of H^+ equivalents is based on the pH sensitivity of the emission of fluorescein in liposomes and proteoliposomes containing traces of the fluorescein-labeled phospholipid DHPE-Fl [41, 42].

   (a) Reconstitute purified Cx26 in a mix of lipids containing PC, PS, and DHPE-Fl at a PC:PS:DHPE-Fl ratio of 2:1:0.005 (w/w/w).

   (b) Start with 100 mM KCl, 0.1 mM EGTA, and 25 mM HEPES/KOH, pH 7.6, and reduce extraliposomal pH to

6.4 by HCl addition to the buffer above, and monitor the decrease in intraliposomal pH by the quenching of the fluorescein bound to the inner leaflet of the liposomes' bilayer (Fig. 4b).

(c) Use the detection conditions described for Fluo-5 N.

(d) Calculate the transport of H^+ equivalents from the change in protonated [HEPES] upon acidification, with protonated [HEPES] calculated from the Henderson–Hasselbalch equation using a pKa of 7.55 [41, 42].

7. Analysis of the changes in fluorescence. Fit the changes in fluorescence in the stopped-flow experiments to multiexponential functions. Use fits to three-exponential functions with the Applied Photophysics Pro-Data Viewer analysis software or equivalent and assess the goodness of fit from the random residual distribution, which should show no structure and chi-square values near unity. Calculate the initial rates of change in concentrations for the 0- to 10-ms period using the faster exponential rate, expressed over a 1-s period [41].

4 Notes

1. Be sure to select true white colonies for analysis. True white colonies tend to be large, so choose the largest and most isolated white colonies. Avoid picking colonies that appear gray or are darker in the center, as they can contain a mixture of cells with empty and recombinant bacmid.
2. To confirm the correct selection of the white colonies, we usually pick 10 colonies and re-streak them on fresh LB agar plates containing 50 μg/mL kanamycin, 7 μg/mL gentamicin, 10 μg/mL tetracycline, 100 μg/mL Bluo-gal, and 40 μg/mL IPTG, and evaluate them after incubation overnight at 37 °C.
3. Bacmid DNA can be stored at −20 °C, aliquoted into separate tubes in TE buffer (10 mM Tris/HCl, 0.1 mM EDTA, pH 8.0). Be sure to avoid freeze/thaw cycles as they decrease transfection efficiency. You may also store the purified bacmid DNA for up to 2 weeks at 4 °C.
4. Budded virus should be released into the medium 72 h after transfection. However, if transfection efficiency is not optimal, cells may not show all the signs of viral infection until 4–5 days post-transfection.
5. Store viral stocks at 4 °C protected from light. Add fetal bovine serum to a final concentration of 2% if the medium is serum-free (e.g., Sf-900 II SFM or Sf-900 III SFM).

6. The mix of lipids can be briefly sonicated if it does not become clear.
7. It is important to prewet the extruder parts by passing a syringe full of buffer through the extruder to reduce dead volume.
8. Do not apply excessive pressure to the syringe to pass the suspension through the membrane, as the membrane can break. It is better to apply a constant low pressure.
9. The method of reconstitution can be chosen depending on the final application, as reconstitution by dialysis and extrusion yields ~50% inside-out Cx26 HCs [41] and 100% inside-out Cx43 HCs [34], whereas gel filtration followed by extrusion yields 100% inside-out Cx26 HCs [42]. The reasons for the differences are unknown.
10. Loading the liposomes and proteoliposomes by extrusion was less efficient, but the functional data were similar, indicating that the HCs were not affected by the freeze-thaw procedure.

Acknowledgments

The implementation and development of the methods presented here were supported in part by grants from the National Institutes of Health (R01GM79629, 3R01GM079629-03S1, and R21DC007150), American Heart Association, Texas Affiliate Inc. (14GRNT18750014 and 0755002Y), and Texas Advanced Research Program (010674-0046-2007).

References

1. Abascal F, Zardoya R (2013) Evolutionary analyses of gap junction protein families. Biochim Biophys Acta 1828(1):4–14. https://doi.org/10.1016/j.bbamem.2012.02.007
2. Mikalsen SO, i Kongsstovu S, Tausen M (2021, 2021) Connexins during 500 million years—from cyclostomes to mammals. Int J Mol Sci 22(4). https://doi.org/10.3390/ijms22041584
3. Mese G, Richard G, White TW (2007) Gap junctions: basic structure and function. J Invest Dermatol 127(11):2516–2524. https://doi.org/10.1038/sj.jid.5700770
4. Nielsen MS, Nygaard Axelsen L, Sorgen PL, Verma V, Delmar M, Holstein-Rathlou NH (2012) Gap junctions. Compr Physiol 2(3):1981–2035. https://doi.org/10.1002/cphy.c110051
5. Harris AL, Locke D (2009) Permeability of connexin channels. In: Harris AL, Locke D (eds) Connexins: A guide. Humana Press, New York
6. Valiunas V, Cohen IS, Brink PR (2018) Defining the factors that affect solute permeation of gap junction channels. Biochim Biophys Acta Biomembr 1860(1):96–101. https://doi.org/10.1016/j.bbamem.2017.07.002
7. Vinken M (2015) Introduction: connexins, pannexins and their channels as gatekeepers of organ physiology. Cell Mol Life Sci 72(15):2775–2778. https://doi.org/10.1007/s00018-015-1958-3
8. Retamal MA, Reyes EP, Garcia IE, Pinto B, Martinez AD, Gonzalez C (2015) Diseases associated with leaky hemichannels. Front Cell Neurosci 9:267. https://doi.org/10.3389/fncel.2015.00267
9. Giaume C, Leybaert L, Naus CC, Saez JC (2013) Connexin and pannexin hemichannels in brain glial cells: properties, pharmacology, and roles. Front Pharmacol 4:88. https://doi.org/10.3389/fphar.2013.00088
10. Giaume C, Naus CC, Saez JC, Leybaert L (2021) Glial connexins and pannexins in the

healthy and diseased brain. Physiol Rev 101(1): 93–145. https://doi.org/10.1152/physrev.00043.2018

11. Wang N, De Bock M, Decrock E, Bol M, Gadicherla A, Vinken M et al (2013) Paracrine signaling through plasma membrane hemichannels. Biochim Biophys Acta Biomembr 1828(1):35–50. https://doi.org/10.1016/j.bbamem.2012.07.002
12. Srinivas M, Verselis VK, White TW (2018) Human diseases associated with connexin mutations. Biochim Biophys Acta Biomembr 1860(1):192–201. https://doi.org/10.1016/j.bbamem.2017.04.024
13. Cocozzelli AG, White TW (2019) Connexin 43 mutations lead to increased hemichannel functionality in skin disease. Int J Mol Sci 20(24). https://doi.org/10.3390/ijms20246186
14. Saez JC, Retamal MA, Basilio D, Bukauskas FF, Bennett MV (2005) Connexin-based gap junction hemichannels: gating mechanisms. Biochim Biophys Acta Biomebr 1711(2): 215–224. https://doi.org/10.1016/j.bbamem.2005.01.014
15. Fiori MC, Reuss L, Cuello LG, Altenberg GA (2014) Functional analysis and regulation of purified connexin hemichannels. Front Physiol 5:71. https://doi.org/10.3389/fphys.2014.00071
16. Sanchez HA, Orellana JA, Verselis VK, Saez JC (2009) Metabolic inhibition increases activity of connexin-32 hemichannels permeable to Ca^{2+} in transfected HeLa cells. Am J Physiol Cell Physiol 297(3):C665–C678. https://doi.org/10.1152/ajpcell.00200.2009
17. Retamal MA, Schalper KA, Shoji KF, Bennett MV, Saez JC (2007) Opening of connexin 43 hemichannels is increased by lowering intracellular redox potential. Proc Natl Acad Sci USA 104(20):8322–8327. https://doi.org/10.1073/pnas.0702456104
18. Retamal MA, Bennett MV, Pelegrin P, Fernandez R (2016) Ion channels in inflammatory processes: what is known and what is next? Mediat Inflamm 2016:6245731. https://doi.org/10.1155/2016/6245731
19. Schulz R, Gorge PM, Gorbe A, Ferdinandy P, Lampe PD, Leybaert L (2015) Connexin 43 is an emerging therapeutic target in ischemia/reperfusion injury, cardioprotection and neuroprotection. Pharmacol Ther 153:90–106. https://doi.org/10.1016/j.pharmthera.2015.06.005
20. Bargiotas P, Monyer H, Schwaninger M (2009) Hemichannels in cerebral ischemia. Curr Mol Med 9(2):186–194
21. Sanchez HA, Villone K, Srinivas M, Verselis VK (2013) The D50N mutation and syndromic deafness: altered Cx26 hemichannel properties caused by effects on the pore and intersubunit interactions. J Gen Physiol 142(1):3–22. https://doi.org/10.1085/jgp.201310962
22. Sanchez HA, Mese G, Srinivas M, White TW, Verselis VK (2010) Differentially altered Ca^{2+} regulation and Ca^{2+} permeability in Cx26 hemichannels formed by the A40V and G45E mutations that cause keratitis ichthyosis deafness syndrome. J Gen Physiol 136(1):47–62. https://doi.org/10.1085/jgp.201010433
23. Sanchez HA, Slavi N, Srinivas M, Verselis VK (2016) Syndromic deafness mutations at Asn 14 differentially alter the open stability of Cx26 hemichannels. J Gen Physiol 148(1):25–42. https://doi.org/10.1085/jgp.201611585
24. Shuja Z, Li L, Gupta S, Mese G, White TW (2016) Connexin26 mutations causing palmoplantar keratoderma and deafness interact with connexin43, modifying gap junction and hemichannel properties. J Invest Dermatol 136(1): 225–235. https://doi.org/10.1038/JID.2015.389
25. Garcia IE, Maripillan J, Jara O, Ceriani R, Palacios-Munoz A, Ramachandran J et al (2015) Keratitis-ichthyosis-deafness syndrome-associated Cx26 mutants produce nonfunctional gap junctions but hyperactive hemichannels when co-expressed with wild type Cx43. J Invest Dermatol 135(5): 1338–1347. https://doi.org/10.1038/jid.2015.20
26. Peng B, Xu C, Wang S, Zhang Y, Li W (2022) The role of connexin hemichannels in inflammatory diseases. Biology (Basel) 11(2). https://doi.org/10.3390/biology11020237
27. Zhang Y, Khan S, Liu Y, Siddique R, Zhang R, Yong VW et al (2021) Gap junctions and hemichannels composed of connexins and pannexins mediate the secondary brain injury following intracerebral hemorrhage. Biology (Basel) 11(1). https://doi.org/10.3390/biology11010027
28. Garcia-Rodriguez C, Bravo-Tobar ID, Duarte Y, Barrio LC, Saez JC (2022) Contribution of non-selective membrane channels and receptors in epilepsy. Pharmacol Ther 231:107980. https://doi.org/10.1016/j.pharmthera.2021.107980
29. Retamal MA, Fernandez-Olivares A, Stehberg J (2021) Over-activated hemichannels: a possible therapeutic target for human diseases. Biochim Biophys Acta Mol basis Dis 1867(11): 166232. https://doi.org/10.1016/j.bbadis.2021.166232

30. Chen Y, Deng Y, Bao X, Reuss L, Altenberg GA (2005) Mechanism of the defect in gap-junctional communication by expression of a connexin 26 mutant associated with dominant deafness. FASEB J 19(11):1516–1518. https://doi.org/10.1096/fj.04-3491fje
31. Lampe PD, Lau AF (2000) Regulation of gap junctions by phosphorylation of connexins. Arch Biochem Biophys 384(2):205–215. https://doi.org/10.1006/abbi.2000.2131
32. Delmar M, Coombs W, Sorgen P, Duffy HS, Taffet SM (2004) Structural bases for the chemical regulation of connexin43 channels. Cardiovasc Res 62(2):268–275. https://doi.org/10.1016/j.cardiores.2003.12.030
33. Bao X, Altenberg GA, Reuss L (2004) Mechanism of regulation of the gap junction protein connexin 43 by protein kinase C-mediated phosphorylation. Am J Physiol Cell Physiol 286(3):C647–C654. https://doi.org/10.1152/ajpcell.00295.2003
34. Bao X, Reuss L, Altenberg GA (2004) Regulation of purified and reconstituted connexin 43 hemichannels by protein kinase C-mediated phosphorylation of serine 368. J Biol Chem 279(19):20058–20066. https://doi.org/10.1074/jbc.M311137200
35. Moreno AP, Fishman GI, Spray DC (1992) Phosphorylation shifts unitary conductance and modifies voltage dependent kinetics of human connexin43 gap junction channels. Biophys J 62(1):51–53. https://doi.org/10.1016/S0006-3495(92)81775-7
36. Moreno AP, Saez JC, Fishman GI, Spray DC (1994) Human connexin43 gap junction channels. Regulation of unitary conductances by phosphorylation. Circ Res 74(6):1050–1057. https://doi.org/10.1161/01.res.74.6.1050
37. Takens-Kwak BR, Jongsma HJ (1992) Cardiac gap junctions: three distinct single channel conductances and their modulation by phosphorylating treatments. Pflugers Archiv: Eur J Physiol 422(2):198–200. https://doi.org/10.1007/BF00370421
38. Kwak BR, van Veen TA, Analbers LJ, Jongsma HJ (1995) TPA increases conductance but decreases permeability in neonatal rat cardiomyocyte gap junction channels. Exp Cell Res 220(2):456–463. https://doi.org/10.1006/excr.1995.1337
39. Ek-Vitorin JF, King TJ, Heyman NS, Lampe PD, Burt JM (2006) Selectivity of connexin 43 channels is regulated through protein kinase C-dependent phosphorylation. Circ Res 98(12):1498–1505. https://doi.org/10.1161/01.RES.0000227572.45891.2c
40. Bao X, Lee SC, Reuss L, Altenberg GA (2007) Change in permeant size selectivity by phosphorylation of connexin 43 gap-junctional hemichannels by PKC. Proc Natl Acad Sci USA 104(12):4919–4924. https://doi.org/10.1073/pnas.0603154104
41. Fiori MC, Figueroa V, Zoghbi ME, Saez JC, Reuss L, Altenberg GA (2012) Permeation of calcium through purified connexin 26 hemichannels. J Biol Chem 287(48):40826–40834. https://doi.org/10.1074/jbc.M112.383281
42. Fiori MC, Krishnan S, Cortes DM, Retamal MA, Reuss L, Altenberg GA et al (2015) Functional hemichannels formed by human connexin 26 expressed in bacteria. Biosci Rep 35(2). https://doi.org/10.1042/BSR20140089
43. Scientific TF: Bac-to-bac baculovirus expression system. User guide. An efficient site-specific transposition system to generate baculovirus for high-level expression of recombinant proteins. https://www.thermofisher.com/document-connect/document-connect.html?url=https://assets.thermofisher.com/TFS-Assets%2FLSG%2Fmanuals%2FMAN0000414_BactoBacExpressionSystem_UG.pdf. Accessed 24 May 2022
44. Scientific TF: Guide to baculovirus expression vector systems (BEVS) and insect cell culture Techniques. https://www.thermofisher.com/document-connect/document-connect.html?url=https://assets.thermofisher.com/TFS-Assets%2FLSG%2Fmanuals%2Fbevtest.pdf. Accessed 24 May 2022
45. Harris AL, Walter A, Zimmerberg J (1989) Transport-specific isolation of large channels reconstituted into lipid vesicles. J Membr Biol 109(3):243–250. https://doi.org/10.1007/BF01870281
46. Kim DY, Kam Y, Koo SK, Joe CO (1999) Gating connexin 43 channels reconstituted in lipid vesicles by mitogen-activated protein kinase phosphorylation. J Biol Chem 274(9):5581–5587. https://doi.org/10.1074/jbc.274.9.5581
47. Natha CM, Vemulapalli V, Fiori MC, Chang CT, Altenberg GA (2021) Connexin hemichannel inhibitors with a focus on aminoglycosides. Biochim Biophys Acta Mol basis Dis 1867(6):166115. https://doi.org/10.1016/j.bbadis.2021.166115

Chapter 2

Expression and Function of Connexins in Extracellular Vesicles

Tânia Martins-Marques and Henrique Girão

Abstract

Extracellular vesicles (EVs) are recognized as major vehicles for exchange of information across distant cells and tissues, which have been extensively explored for diagnosis and therapeutic purposes. The presence of multiple connexin (Cx) proteins has been described in EVs, where they might facilitate EV–cell communication. However, quantitative changes in Cx levels and functional assessment of Cx channels have only been established for Cx43. In present work, we provide a detailed description of the protocols we have optimized to assess the expression and permeability of Cx43 channels in EVs derived from cultured cells and human peripheral blood. Particularly, we include some modifications to improve quantitative analysis of EV-Cx43 by enzyme-linked immunosorbent assay (ELISA) and assessment of channel functionality by sucrose-density gradient ultracentrifugation, which can be easily adapted to other Cx family members, leveraging the development of diagnostic and therapeutic applications based on Cx-containing EVs.

Key words Connexins, Connexin43, Extracellular vesicles, Vesicles isolation, Hemichannels, Transport-specific fractionation assay, Enzyme-linked immunosorbent assay, Size-exclusion chromatography, Sucrose-density gradient ultracentrifugation

1 Introduction

Secreted by almost every cell type, extracellular vesicles (EVs) represent a heterogeneous population of nanosized vesicles enclosing a large variety of active biomolecules, including proteins, lipids, and genetic material [1, 2]. Since EVs are ubiquitously present in biological fluids, changes in the number and content of EVs can serve as disease biomarkers [3–5]. The presence of several gap junction (GJ) proteins, including connexin (Cx)26, Cx32, Cx43, Cx45, and Cx46, has been reported in EVs secreted by a wide variety of cells [6–8]. Mechanistic insights were mainly obtained using Cx43 as a model, which allowed the identification of functional hexameric Cx43 channels at the EV surface that facilitate rapid unloading of vesicle contents [9]. Given the potential of Cx43-containing EVs to efficiently deliver therapeutic cargo into

Fabio Mammano and Mauricio Retamal (eds.), *Connexin Hemichannels: Methods and Protocols*, Methods in Molecular Biology, vol. 2801, https://doi.org/10.1007/978-1-0716-3842-2_2,

target cells, it is vital to establish robust and reliable tools to assess functionality of Cx channels in EVs samples [10, 11]. Based on a classic method used to evaluate the permeability of liposomal Cx43 hemichannels, we developed a transport-specific fractionation assay to enable functional analysis of Cx43 channels in EVs [12, 13]. This approach relies on the migration profile of EVs in a gradient of sucrose with increased concentration from top to bottom. Briefly, EVs containing permeable Cx43 channels display increased ability to uptake sucrose and migrate to lower positions in the tube (higher sucrose concentration), whereas impermeable EV-Cx43 remain in the upper fractions (lower sucrose concentration) [1]. This method can be applied to study functionality of EVs channels formed by other Cx isoforms, providing a useful platform to study the mechanisms and signals regulating channel gating in EVs.

Disease-associated changes in the levels of Cx43 in EVs circulating in peripheral blood have also been reported, suggesting a role for Cx43-containing EVs as biomarkers of human pathologies [1]. Hence, the development of quantitative measurement techniques is of utmost importance to uncover alterations in Cx43 secretion into EV under pathological conditions, which could be extended to other Cx family members. Here, we describe a double-antibody sandwich enzyme-linked immunosorbent assay (ELISA), in which Cx43-containing EVs are captured by a monoclonal antibody against Cx43, followed by detection with a polyclonal Cx43 antibody and an enzyme-conjugated secondary antibody [14]. After incubation with a chromogenic substrate, an absorbance signal is produced, which can be measured by spectrophotometry. This method allows reliable and high sensitivity quantification, which is particularly useful for detection of Cx43 in vesicles obtained from small amounts of human biological fluids.

2 Materials

2.1 Production of EVs

1. Cell lines (e.g., human embryonic epithelial kidney (HEK)-293 cells, or primary cultures, *see* **Note 1**).
2. Dulbecco's Modified Eagle Medium (DMEM).
3. Phosphate-buffered saline (PBS): 137 mM NaCl, 2.7 mM KCl, 1.5 mM KH_2PO_4, 8.1 mM Na_2HPO_4, pH 7.4.
4. EVs-depleted DMEM: dilute 50% FBS in DMEM and ultracentrifuge at 120,000 × *g*, for 16 h. Carefully pour the supernatant into clean tubes and dilute it to a final concentration of 10% FBS in DMEM.
5. Standard laminar flow cabinet and incubator for cell culture.

2.2 Isolation of EVs by Differential Ultracentrifugation

1. PBS.
2. Refrigerated centrifuge.
3. Ultracentrifuge and fixed-angle or swinging-bucket rotor.
4. Appropriate polypropylene centrifuge tubes and microcentrifuge tubes.
5. 0.22 μm filters and syringes.

2.3 Isolation of EVs by Size-Exclusion Chromatography (SEC)

1. PBS.
2. Amicon Ultra-15 Centrifugal Filter Unit (MWCO = 100 kDa).
3. Sepharose CL-4B.
4. Plastic syringe (10 mL, or other volumes, as appropriate).
5. Three-way stopcock.
6. Nylon stocking (20 denier).
7. Refrigerated centrifuge.
8. Microcentrifuge tubes.
9. Support stand with clamps.

2.4 Analysis of Permeability of Cx43-Containing EVs by Transport-Specific Fractionation Assay

1. PBS.
2. Sucrose solutions: 0.4 M, 2 M, and 2.5 M. Dilute in ultrapure water and adjust pH to 7.4.
3. Appropriate polypropylene centrifuge tubes and microcentrifuge tubes.
4. Ultracentrifuge and fixed-angle or swinging-bucket rotor.
5. Atago Uricon-N Hand-held Refractometer.

2.5 Analysis of Cx43 Expression in EVs by Immunoblot

1. RIPA buffer: 150 mM NaCl, 50 mM Tris–HCl, 1% NP-40, 0.1% SDS pH 7.4, supplemented with protease inhibitors (e.g., protease inhibitor cocktail (Roche), 2 mM phenylmethanesulfonyl fluoride (PMSF), 10 mM iodoacetamide, and 2 mM sodium ortovanadate).
2. Colorimetric protein concentration assay.
3. Laemmli sample buffer (4×): 250 mM Tris–HCl pH 6.8, 8% (w/v) SDS, 40% (v/v) glycerol, 20% (v/v) β-mercaptoethanol (optional), 0.012% bromophenol blue.
4. Probe sonicator.
5. Refrigerated centrifuge.
6. Reagents and equipment for sodium dodecyl sulfate-polyacrylamide gel electrophoresis (SDS-PAGE) and immunoblot.
7. Digital imaging system and analysis software.

2.6 Analysis of Cx43 Expression in EVs by ELISA

1. RIPA buffer.
2. Colorimetric protein assay.
3. Recombinant Glutathione S-transferase (GST)-tagged Cx43 soluble carboxyl-terminus.
4. Anti-Cx43 capture antibody (#610062, BD Biosciences).
5. Anti-Cx43 detection antibody (#710700, Thermo Fisher Scientific).
6. Anti-rabbit horseradish peroxidase (HRP)-conjugated secondary antibody.
7. Wash buffer: PBS with 0.05% Tween 20.
8. Blocking buffer: 1% bovine serum albumin (BSA) solution in PBS.
9. Substrate solution: 3,3′, 5,5-tetramethylbenzidine (TMB).
10. Stop solution: 1 N HCl.
11. Polystyrene microplates for ELISA.
12. Microplate reader.

3 Methods

3.1 Production of EVs

In this section, we describe in detail the procedure for isolation of EVs from conditioned medium of cultured cells or human plasma/serum samples (*see* **Note 1**). Although several methods have been described for the purification of EVs from different biological fluids [15], we will be focused on differential ultracentrifugation and SEC (*see* **Notes 2–4**).

1. Grow cells in appropriate cell culture medium, until 70–90% confluency is reached.
2. Wash the cells with PBS and replace the medium for an EVs-depleted medium, for 24–48 h.
3. Harvest the conditioned culture medium, prior to the isolation of EVs, following one of the methods described in Subheadings 3.2 or 3.3.

3.2 Isolation of EVs by Differential Ultracentrifugation

1. Centrifuge conditioned medium at 300 × *g*, for 10 min, to remove dead cells and cellular debris.
2. Pour the supernatant into clean centrifuge tubes and centrifuge at 16,500 × *g*, for 20 min.
3. Collect the supernatant and pass it through a 0.22 μm filter unit.
4. Ultracentrifuge the filtrates at 120,000 × *g*, for 70 min.

5. Resuspend the EVs pellets in an appropriate buffer solution (*see* **Note 5**).

3.3 Isolation of EVs by SEC

1. Load conditioned cell culture medium onto the centrifugal filter units (MWCO = 100 kDa).
2. Concentrate culture medium to ~1 mL by repeated centrifugation at 4000 × *g*.
3. Collect concentrated culture medium.
4. Mount a 10 mL plastic syringe that will serve as packing reservoir vertically on a laboratory stand (*see* Fig. 1).
5. Place a nylon stocking on the syringe tip, and a three-way stopcock to control buffer flow through the column.
6. Pack 10 mL of Sepharose CL-4B slurry in the syringe to create a column with a height of approximately 6 cm.
7. Wash and equilibrate the column with 3 volumes of PBS (*see* **Note 6**).
8. Apply 1 mL concentrated cell culture medium (or 1 mL serum/plasma samples) to the equilibrated column (*see* **Note 7**).
9. When the entire sample enters the column, start the sample elution with PBS.
10. Collect 20–30 fractions of 0.5 mL each (*see* **Note 8**).

3.4 Characterization of EVs

In this section, we present some of the methods usually performed to analyze channel permeability of Cx43-enriched EVs samples and Cx43 protein expression. We describe a sucrose-density gradient ultracentrifugation method that enables the assessment of EV-Cx43 channel permeability to small solutes. This approach can be combined with immunoblotting to evaluate the presence of Cx43 and canonical EVs markers in the same fractions. For quantitative assessment of Cx43 levels in purified EVs, double-antibody sandwich ELISA is the most appropriate method.

3.4.1 Transport-Specific Fractionation Assay

1. Equilibrate isolated EVs in 0.4 M sucrose (*see* **Notes 9** and **10**).
2. Layer equilibrated EVs over a discontinuous sucrose step-gradient (0.4–2.5 M—*see* Table 1 and Fig. 2).
3. Ultracentrifuge samples at 160,000 × *g* for 18 h (*see* **Note 11**).
4. Collect sequential fractions (1 mL, from top to bottom).
5. Assess density of all fractions using an Atago Uricon-N Hand-held Refractometer.
6. Wash each fraction with PBS, by ultracentrifugation at 120,000 × *g* for 70 min.
7. Resuspend pelleted EVs in an appropriate buffer solution.

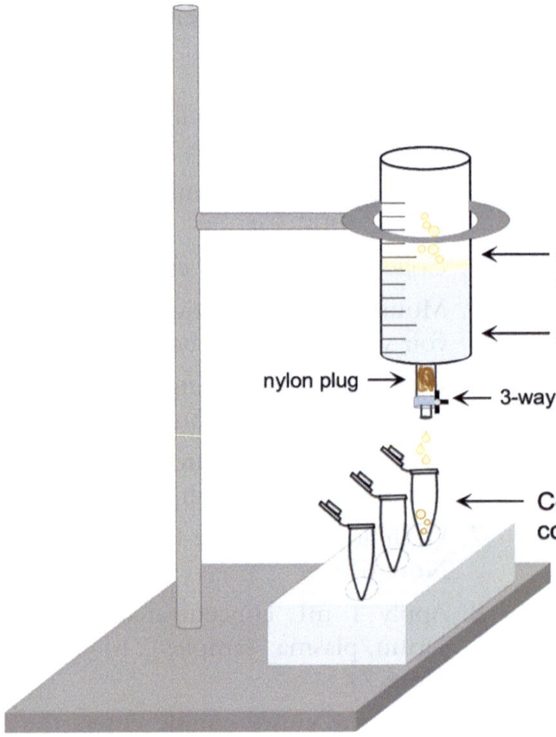

Fig. 1 Schematic representation of the workflow for isolation of extracellular vesicles by size-exclusion chromatography. Concentrated conditioned medium and/or biological fluids containing extracellular vesicles (EVs) are applied to a packed Sepharose column. Sample is eluted with PBS, and fractions are collected for subsequent analysis

3.4.2 Immunoblot

1. Sonicate lysed EVs (*see* **Note 5, list c**) with two 10 s pulses (5 s in between pulses) using a probe sonicator (keep the samples in an ice bath and keep the probe away from the sample-air interface to minimize foaming).
2. Centrifuge the sample at 13,000 × *g*, for 5 min at 4 °C, and transfer the supernatant to a new Eppendorf tube.
3. Measure the total amount of protein, using a commercially available assay. Add denaturing sample buffer, either reducing or nonreducing, as appropriate, and load 20–50 μg of protein/well (*see* **Note 12**).
4. Separate and transfer the proteins by gel electrophoresis and electroblotting.
5. Block and wash the membrane before probing with antibodies against EVs proteins, including Cx43 (*see* **Notes 13** and **14**).
6. Detect the specific protein by chemiluminescence, using a digital imaging system and analysis software.

Table 1
Sucrose molarity in discontinuous gradient fractions of transport-specific fractionation assay

Step-gradient	Sucrose molarity (M)
1	2.500
2	2.000
3	1.886
4	1.771
5	1.657
6	1.543
7	1.429
8	1.314
9	1.200
10	1.086
11	0.971
12	0.857
13	0.743
14	0.629
15	0.514
16	0.400

3.4.3 Elisa

1. Thaw EVs on ice and add SDS (1% final concentration).
2. Denature EVs samples, during 5 min, at 95 °C (*see* **Note 15**).
3. Sonicate and centrifuge lysed EVs during 5 min, at 1200 × *g*.
4. Dilute EVs to 0.1% SDS final concentration.
5. Quantify total protein, using a commercially available assay.
6. Coat ELISA microplate with 100 μL coating solution containing anti-Cx43 capture antibody (dilution 1:25 in blocking buffer; *see* **Note 16**) overnight, at 2–8 °C.
7. Aspirate wells and wash once with 300 μL of wash buffer per well (*see* **Note 17**).
8. Block plate with 200 μL of blocking buffer per well, for 1 h at room temperature.
9. Prepare standards and EVs sample dilutions in blocking buffer.
10. Pipette (in duplicate) 100 μL of optical density standards (20–540 ng/mL of recombinant GST Cx43 soluble carboxyl-terminus, *see* **Note 18**) and EVs samples (5000 ng/μL) into designated wells.

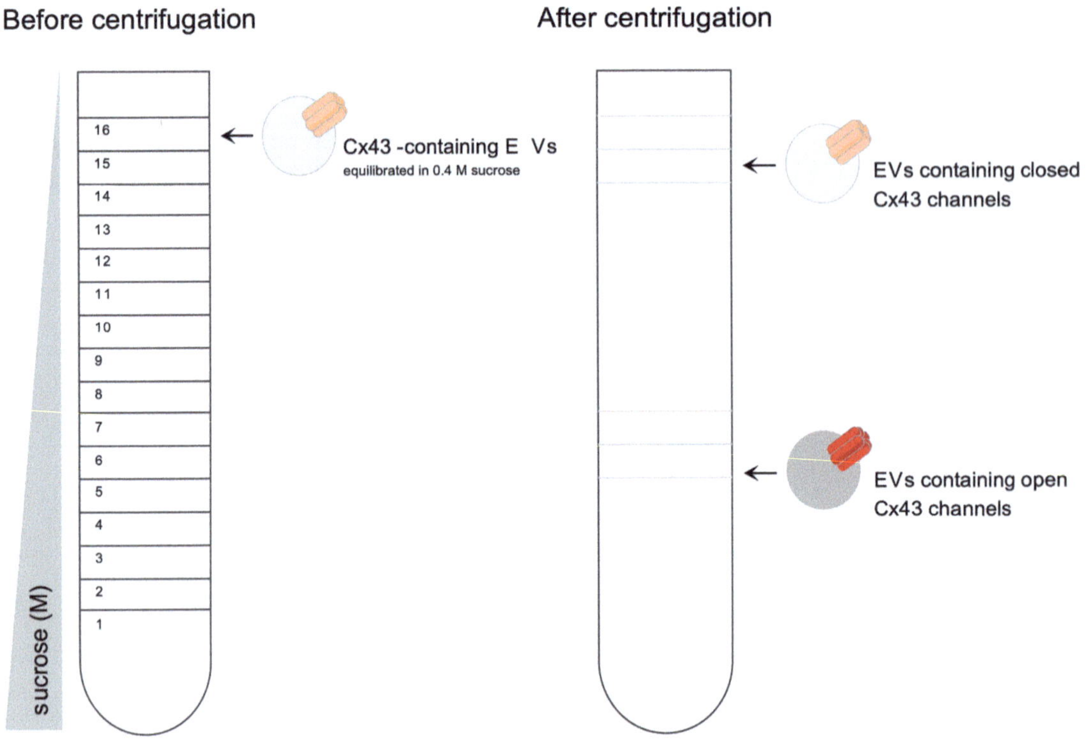

Fig. 2 Schematic representation of the transport-specific fractionation assay. Graphical representation of the discontinuous sucrose step-gradient (0.4–2.5 M) used to assess permeability of Cx43 channels in extracellular vesicles (EVs). Upon centrifugation, EVs containing sucrose-permeable Cx43 channels migrate to lower positions in the tube, whereas sucrose-impermeable EV-Cx43 channels remain in the upper fractions

11. Incubate for 2 h at room temperature.
12. Aspirate wells and wash 5 times with 300 μL of wash buffer per well (*see* **Note 17**).
13. Add 100 μL of anti-Cx43 detection antibody (dilution 1:25 in blocking buffer; *see* **Note 16**) per well, for 2 h at room temperature.
14. Aspirate wells and wash 5 times with 300 μL of wash buffer per well (*see* **Note 17**).
15. Add 100 μL of anti-rabbit HRP antibody (dilution 1:5000) per well, for 20 min at room temperature.
16. Aspirate wells and wash 5 times with 300 μL of wash buffer per well (*see* **Note 17**).
17. Incubate with 100 μL of TMB per well, for 20 min at room temperature.
18. Add 100 μL of stop solution.

19. Measure optical density at 450 nm (background at 570 nm) within 30 min of adding stop solution.
20. Calculate results using a quadratic regression of the calibration curve defined by the optical density of the standards.

4 Notes

1. In addition to EVs isolated from cultured cells, it is possible to purify EVs from serum or plasma samples obtained from blood, withdrawn from human donors or animal models (specific blood collection tubes, with or without anticlotting agents, must be used for each case). To isolate EVs, dilute plasma or serum samples with an equal volume of PBS and centrifuge at 2000 × *g*, for 30 min. Ultracentrifuge the supernatant at 12,000 × *g*, for 45 min. Carefully pour the supernatant and ultracentrifuge at 110,000 × *g*, for 2 h. Resuspend the pellet in 1 mL PBS and transfer it into a clean tube. Dilute the specimen in PBS and filter through a 0.22 μm filter, to remove large vesicles. Ultracentrifuge at 110,000 × *g*, for 70 min. Wash by ultracentrifugation at 110,000 × *g*, for 70 min. Resuspend the EVs pellet in an appropriate buffer.
2. Differential centrifugation methods are gold standards for EVs purification. However, ultracentrifugation at 110,000 × *g* leads to cosedimentation of other contaminants, such as protein aggregates, lipoproteins, and extracellular proteins nonspecifically bound to EVs. Further purification, using sucrose or iodixanol gradients, can be performed to eliminate these contaminants.
3. EVs can also be isolated with commercially available kits, mainly based on precipitation of EVs following the addition of proprietary polymers that mostly restrain water molecules. Major advantages of those strategies are the higher recovery of EVs, particularly from low abundant samples (e.g., biological fluids), quickness, and easiness, since no technical skills or expensive specialized equipment are required. However, coprecipitation of nonvesicular contaminants is also observed.
4. Before western blot (WB) analysis of diluted EVs samples obtained after SEC, total protein can be precipitated to concentrate the sample using standard techniques, namely trichloroacetic acid (TCA) precipitation. Briefly, add 0.11 volumes of ice-cold 100% TCA to each sample, place on ice for 10 min, and add 500 μL of ice-cold 10% TCA. After 20 min on ice, centrifuge at 16,100 × *g* for 30 min. Carefully remove the supernatant and add 500 μL of acetone to the protein pellet, before

centrifugation at 16,100 × *g* for 30 min. Remove the supernatant, let pellet air dry, and resuspend in RIPA buffer.

5. According with the downstream application, EVs pellets should be resuspended in a buffer with appropriate composition. Some examples are the following:
   (a) For nanoparticle tracking analysis (NTA), EVs should be resuspended in particle-free PBS.
   (b) For transmission electron microscopy (TEM), EVs-containing pellets should be resuspended in PBS, followed by fixation in 2% (w/v) paraformaldehyde (PFA).
   (c) For WB analysis, EVs should be resuspended in PBS, followed by the addition of 2x RIPA buffer (or similar).
6. Do not let the SEC column dry, as it will significantly impair its performance. Dry columns must be discarded.
7. It is possible to use different input volumes of conditioned medium or biological fluids. For that, the volume of the column has to be optimized. Typically, for equivalent Sepharose volume, higher column height improves separation of EVs from contaminants. Higher EVs recovery can be obtained in narrower columns with a smaller volume of Sepharose and equivalent height.
8. Each fraction can be analyzed to determine the total amount of particles (e.g., by NTA) and total protein content. In this setup, EVs typically show a peak around fractions 8–12, before the peak of soluble proteins.
9. The transport-specific fractionation assay requires a large amount of EVs. When using EVs derived from cell cultures expressing Cx43, ~3×10^{10} particles should be used. Regarding biological fluids that might present reduced levels of Cx43, typically 5–10 mL serum/plasma should be used for EVs purification before transport-specific fractionation.
10. pH of sucrose solutions must be strictly controlled, as it can interfere with Cx43 channel permeability. For example, EV-Cx43 channel closure is induced by the use of sucrose solutions with low pH levels (pH 6.0).
11. To prevent gradient disturbance, acceleration and deceleration steps of the ultracentrifugation should be made at slower rates.
12. Nonreducing conditions (i.e., classical Laemmli sample buffer without DTT or β-mercaptoethanol) should be used to assess the oligomerization state of Cx43 (monomeric form and/or hexameric channels) in EVs. For detection of CD9, CD63, CD81, and ICAM-1, nonreducing conditions should also be

employed, as the disulfide bond on the epitope of those antigens is often recognized by the antibodies.

13. Protein content of EVs should be characterized by the presence of transmembrane or GPI-anchored proteins associated to plasma membrane and/or endosomes (e.g., tetraspanins CD63, CD81, CD9, or MHC class I and II), and cytosolic proteins recovered in EVs (e.g., TSG101, ALIX (PDCD6IP), Flotillins-1 and 2, or Hsc70). To ensure purity of isolated EVs, analysis of most common coisolated contaminants (e.g., lipoproteins or ribosomal proteins) should be performed [15].
14. WB analysis of purified EVs should be performed by loading side-by-side EVs samples and lysates of vesicle-producing cells. Equivalent amounts of total protein should be loaded to determine enrichment of the analyzed proteins in EVs. When analyzing EVs from biological fluids, whole cell lysates from the same species can be used as technical control.
15. Heat denaturation of the samples is required to minimize antigen conformation variability, ensuring a more reliable and reproducible quantification of Cx43 levels in EVs.
16. Antibody concentration (capture and detection antibodies) should be optimized to determine the dynamic range of detection. Serial antibody dilutions should be performed to titrate antibody concentration.
17. To improve reproducibility in the washing procedure, invert plate and tap on absorbent paper to remove excess liquid.
18. Recombinant GST-Cx43 carboxyl-terminus can be purchased from a biological products company (e.g., MyBiosource, San Diego, CA, USA) or generated by cloning the appropriate cDNA into a pGEX4T1 vector. Briefly, recombinant protein can be expressed in *Escherichia coli* BL21, induced by 0.1 mM isopropyl β-D-1-thiogalactopyranoside (IPTG) overnight at 20 °C, and purified using GSTrap™ high-performance columns (GE Healthcare).

Acknowledgments

This work was supported by the European Regional Development Fund (ERDF) through the Operational Program for Competitiveness Factors (COMPETE) (under the projects HealthyAging2020 CENTRO-01-0145-FEDER-000012-N2323; CENTRO-01-0145-FEDER-032179, CENTRO-01-0145-FEDER-032414, POCI-01-0145-FEDER-022122, UIDB/04539/2020, and UIDP/04539/2020) and EU-CARDIOPROTECTION COST Action (CA16225), supported by COST (European Cooperation in Science and Technology).

References

1. Martins-Marques T, Ribeiro-Rodrigues T, de Jager SC et al (2020) Myocardial infarction affects Cx43 content of extracellular vesicles secreted by cardiomyocytes. Life Sci Alliance 3:e202000821
2. Lee H, Li C, Zhang Y et al (2019) Caveolin-1 selectively regulates microRNA sorting into microvesicles after noxious stimuli. J Exp Med 216:2202–2220
3. Li M, Zeringer E, Barta T et al (2014) Analysis of the RNA content of the exosomes derived from blood serum and urine and its potential as biomarkers. Philos Trans R Soc B: Biol Sci 369: 654–659
4. Colombo M, Raposo G, Théry C (2014) Biogenesis, secretion, and intercellular interactions of exosomes and other extracellular vesicles. Annu Rev Cell Dev Biol 30:255–289
5. Turay D, Khan S, Diaz Osterman CJ et al (2016) Proteomic profiling of serum-derived exosomes from ethnically diverse prostate cancer patients. Cancer Investig 34:1–11
6. Ribeiro-Rodrigues TM, Martins-Marques T, Morel S et al (2017) Role of connexin 43 in different forms of intercellular communication—gap junctions, extracellular vesicles and tunnelling nanotubes. J Cell Sci 130:3619–3630
7. Falleni A, Moscato S, Sabbatini ARM et al (2021) Subcellular localization of connexin 26 in cardiomyocytes and in cardiomyocyte-derived extracellular vesicles. Molecules 26: 6726
8. Acuña RA, Varas-Godoy M, Berthoud VM et al (2020) Connexin-46 contained in extracellular vesicles enhance malignancy features in breast cancer cells. Biomol Ther 10:676
9. Soares AR, Martins-Marques T, Ribeiro-Rodrigues T et al (2015) Gap junctional protein Cx43 is involved in the communication between extracellular vesicles and mammalian cells. Sci Rep 5:13243
10. Martins-Marques T, Pinho MJ, Zuzarte M et al (2016) Presence of Cx43 in extracellular vesicles reduces the cardiotoxicity of the anti-tumour therapeutic approach with doxorubicin. J Extracell Vesicles. 5:32538
11. Kojima R, Bojar D, Rizzi G et al (2018) Designer exosomes produced by implanted cells intracerebrally deliver therapeutic cargo for Parkinson's disease treatment. Nat Commun 9:1305
12. Bao X, Reuss L, Altenberg GA (2004) Regulation of purified and reconstituted Connexin 43 Hemichannels by protein kinase C-mediated phosphorylation of serine 368. J Biol Chem 279:20058–20066
13. Fiori MC, Reuss L, Cuello LG et al (2014) Functional analysis and regulation of purified connexin hemichannels. Front Physiol 5:71
14. Logozzi M, Di Raimo R, Mizzoni D et al (2020) Immunocapture-based ELISA to characterize and quantify exosomes in both cell culture supernatants and body fluids. Methods Enzymol 645:155
15. Théry C, Witwer KW, Aikawa E et al (2018) Minimal information for studies of extracellular vesicles 2018 (MISEV2018): a position statement of the international society for extracellular vesicles and update of the MISEV2014 guidelines. J Extracell Vesicles 7:1535750

Chapter 3

Expression, Purification, and Nanodisc Reconstitution of Connexin-43 Hemichannels for Structural Characterization by Cryo-Electron Microscopy

Pia Lavriha, Chao Qi, and Volodymyr M. Korkhov

Abstract

Connexins are polytopic domain membrane proteins that form hexameric hemichannels (HCs) which can assemble into gap junction channels (GJCs) at the interface of two neighboring cells. The HCs may be involved in ion and small-molecule transport across the cellular plasma membrane in response to various stimuli. Despite their importance, relatively few structures of connexin HCs are available to date, compared to the structures of the GJCs. Here, we describe a protocol for expression, purification, and nanodisc reconstitution of connexin-43 (Cx43) HCs, which we have recently structurally characterized using cryo-EM analysis. Application of similar protocols to other connexin family members will lead to breakthroughs in the understanding of the structure and function of connexin HCs.

Key words Connexin-43, Hemichannel, Gap junction channel, Reconstitution, Nanodisc, Cryo-EM, Single-particle analysis

1 Introduction

Connexins are the building blocks of the gap junction channels (GJCs), which mediate the metabolic and electric coupling of a wide range of tissues [1]. Connexin monomers assemble into hexameric hemichannels (HCs). The HCs at the plasma membrane are able to interact with the HCs on the surface of a neighboring cell, leading to formation of the dodecameric gap junction channels (GJCs) spanning two adjacent lipid bilayers and typically organized into two-dimensional arrays [2]. GJCs facilitate the passage cellular components (e.g., ions or metabolites) below the molecular weight of ~1500 Da [3].

The Authors Pia Lavriha and Chao Qi have contributed equally to this chapter

Fabio Mammano and Mauricio Retamal (eds.), *Connexin Hemichannels: Methods and Protocols*, Methods in Molecular Biology, vol. 2801, https://doi.org/10.1007/978-1-0716-3842-2_3,

One of the most extensively studied connexin genes is connexin-43 (Cx43, *GJA1*), which was discovered as the major protein in the heart gap junctions [4]. The Cx43 GJCs are critical for normal tissue homeostasis, mediating the propagation of action potentials in the heart [5], participating in the neuroglial syncytium maintenance [6], etc. Moreover, the HCs of Cx43 also have important physiological functions: for example, the HCs have been shown to mediate the release of ATP in a variety of cell types [7, 8]. Several genetic diseases are associated with mutations in Cx43: oculodentodigital dysplasia [9–11], Hallermann–Streiff syndrome [12], hypoplastic left heart syndrome 1 [13], atrioventricular septal defect 3 [13], and others. The precise role of the HCs or GJCs in these conditions is not fully established. Nevertheless, the medical importance of HCs is well established and has led to drug discovery efforts targeting the HCs in cancer, inflammation, or skin wounds [3].

Structural studies of connexin GJCs have a rich history, from cryo-electron microscopy (cryo-EM) of 2D crystals of Cx43 at low resolution in the 1990s [14], to X-ray crystallography (X-ray structure of Cx26 GJC) [15] and, more recently, single-particle cryo-EM-based studies (such as the structures of Cx46/50 GJC) [16, 17]. In contrast, relatively few studies successfully tackled the structures of connexin HCs, possibly due to the difficulties in stabilizing the HCs or due to their propensity to form GJCs in vitro. The few examples are: the cryo-EM structure of Cx31.3 HC [18] and the structure of the Cx26 mutant that prevents GJC formation [19]. We have recently succeeded in determining the high-resolution structures of Cx43 GJC in detergent and in nanodisc [20]. Interestingly, the nanodisc-reconstituted sample also contained a substantial proportion of Cx43 HC, which allowed us to determine the cryo-EM structure of the HC at ~4 Å resolution.

Nanodisc reconstitution is a powerful tool for preparing in vitro membrane protein samples for downstream biochemical, biophysical, or structural studies. The method relies on the ability of the membrane proteins purified in detergent to integrate into the lipidic environment of the nanodisc, the membrane scaffold protein (MSP)-delimited lipid-protein nanoparticles, upon detergent removal (reviewed in [21]). Several MSPs have been developed based on apolipoprotein A-1, tailored to incorporate membrane proteins of different sizes. In the case of Cx43 channels, we opted to use MSP2N2, a high lipid capacity/large size nanodisc-forming MSP [22] (Fig. 1). This MSP is likely to provide suitable nanodisc environment for other connexin channels.

This chapter describes the detailed protocol for Cx43 expression, purification, and nanodisc reconstitution which will be helpful in the future structure-function studies of other connexin HCs.

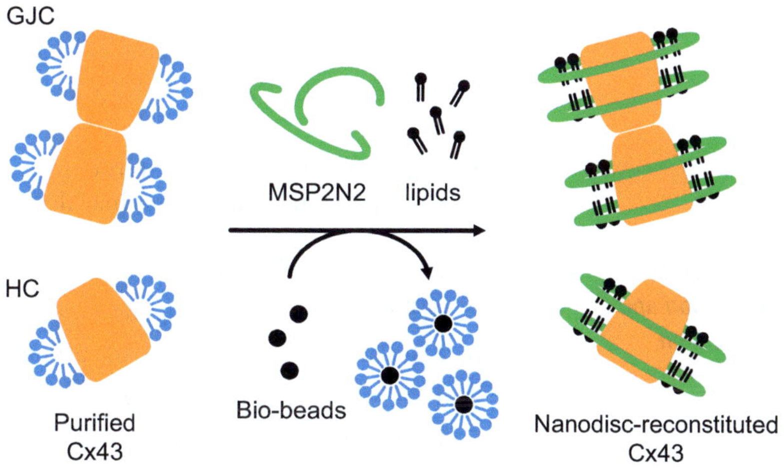

Fig. 1 Reconstitution of the Cx43 channels into nanodiscs. The purified Cx43 channels (GJCs and HCs) are reconstituted using a procedure that involves mixing of the protein with the lipids and the membrane scaffold protein (MSP2N2 in this case), followed by the addition of the Bio-beads for detergent removal. In the absence of the detergent micelles surrounding the GJCs and HCs, the proteins integrate into the lipid particles formed by the MSP2N2 proteins. The nanodiscs embedding the membrane proteins in a lipid environment can be used for the downstream biochemical, biophysical, or structural characterization

2 Materials

All solutions should be prepared in milliQ water (18 MΩ-cm) using analytical grade chemicals. The detergent for solubilization is sol-grade.

2.1 Cx43 Protein Expression in Mammalian Cells

1. Dulbecco's Modified Eagle's Medium with High Glucose (4.5 g/L) and L-glutamine, supplemented with 10% fetal calf serum (FCS) and penicillin/streptomycin (PenStrep) for cell culture.
2. Dulbecco's Modified Eagle's Medium with High Glucose (4.5 g/L) and L-glutamine, supplemented with 2% fetal calf serum (FCS) and penicillin/streptomycin (PenStrep) for cell culture.
3. Dulbecco's Modified Eagle's Medium with High Glucose (4.5 g/L) and L-glutamine.
4. Branched polyethyleneimine (PEI 25 kDa) can be prepared as 100 mg/mL stock in dH_2O and diluted to a 1 mg/mL concentration in dH_2O and sterile filtered through a 0.20 μm filter.
5. Basic cell culture plastic ware (15 cm sterile petri dishes).

6. An expression vector containing the human Cx43 (Uniprot ID: P17302) with a C-terminal 3C-EYFP-twinStrep tag.
7. Freestyle 293-F cells (HEK293F; Thermo Fisher Scientific) (*see* **Note 1**).
8. Appropriate space for sterile mammalian cell culture work equipped with a BSL1 laminar flow.

2.2 MSP2N2 Expression and Purification

1. *E. coli* BL21[DE3] cells.
2. *MSP2N2* plasmid (Addgene #29520).
3. LB plates supplemented with 50 μg/mL kanamycin.
4. 1 M isopropyl β-D-1-thiogalactopyranoside (IPTG).
5. Bacterial cell shaking incubator.
6. Large centrifuge (i.e., Herolab HiCen XL equipped with F8S-6x1000Y rotor).
7. Lysis buffer: 40 mM Tris-HCl pH 8.0, 0.3 M NaCl, 1% Triton X-100.
8. Wash buffer 1: 40 mM Tris-HCl pH 8.0, 0.3 M NaCl, 50 mM Na-cholate, 20 mM imidazole.
9. Wash buffer 2: 40 mM Tris-HCl pH 8.0, 0.3 M NaCl, 50 mM imidazole.
10. Elution buffer: 40 mM Tris-HCl pH 8.0, 0.3 M NaCl, 0.4 M imidazole.
11. Desalting buffer: 20 mM Tris-HCl pH 8.0, 0.1 M NaCl, 0.5 mM EDTA.
12. DNase I (Sigma-Aldrich #10104159001).
13. Ni-NTA resin (Thermo Fischer Scientific #88222).
14. Amicon Ultra concentrator, 10 kDa cut-off.
15. G-25 PD 10 desalting column (Cytiva #17085101).

2.3 GFP Nanobody Expression and Purification

1. *E. coli* BL21[DE3] cells.
2. A bacterial expression vector containing the anti-GFP nanobody sequence with a His-tag (Addgene, plasmid #49172) [23].
3. LB plates supplemented with 100 μg/mL ampicillin.
4. Complete TB medium: 1× TB salts, 1% glucose, appropriate antibiotic (*see* **Note 2**).
5. 1 M isopropyl β-D-1-thiogalactopyranoside (IPTG).
6. Bacterial cell shaking incubator.
7. Large centrifuge (i.e., Herolab HiCen XL equipped with F8S-6x1000Y rotor).

8. Resuspension buffer: 20 mM HEPES pH 7.5, 150 mM NaCl, 1 mM PMSF.
9. Wash buffer 1: 20 mM HEPES pH 7.5, 150 mM NaCl, 20 mM imidazole.
10. Wash buffer 2: 20 mM HEPES pH 7.5, 150 mM NaCl, 40 mM imidazole.
11. Elution buffer: 20 mM HEPES pH 7.5, 150 mM NaCl, 250 mM imidazole.
12. SEC buffer: 20 mM HEPES pH 7.5, 150 mM NaCl.
13. Microfluidizer (i.e., LM10 Microfluidizer) (*see* **Note 3**).
14. Ni-NTA resin (Thermo Scientific #88222).
15. Superdex 75 16/600 GL size-exclusion chromatography column or equivalent.
16. Amicon Ultra concentrator, 10 kDa cut-off.
17. Nanodrop spectrometer.
18. Glycerol.

2.4 Coupling of CNBr-Sepharose with Anti-GFP Nanobody

1. 1 g of freeze-dried CNBr-Sepharose 4B beads (Cytiva #17043001) (*see* **Note 4**).
2. Purified GFP nanobody in 20 mM HEPES pH 7.5, 150 mM NaCl, 10% glycerol.
3. 1 mM HCl.
4. 0.1 M $NaHCO_3$ pH 8.5, 0.5 M NaCl.
5. 0.2 M Glycine-NaOH pH 8.0.
6. Buffer A: 25 mM Tris-HCl, pH 8.0, 150 mM NaCl.
7. Glass gravity column (Bio-Rad Econo-Columns, 2.5 × 10 cm).

2.5 Cx43 Protein Purification

1. Buffer A: 25 mM Tris-HCl, pH 8.0, 150 mM NaCl.
2. Buffer B: 25 mM Tris–HCl, pH 8.0, 150 mM NaCl, 0.1% digitonin (w/v). Prepare the 5% (w/v) stock solution 2 days prior to purification. To prepare the 5% stock solution, dissolve the digitonin powder in a falcon tube in heated Mili-Q water. Vortex the falcon tube briefly and place in a water bath at 95 °C for 30 min or until the powder is fully dissolved. Place the tube at 4 °C overnight. Filter the digitonin stock prior to use.
3. Protease inhibitors: 1 mM benzamidine, 1 μg/mL leupeptin, 1 μg/mL aprotinin, 1 μg/mL pepstatin, 1 μg/mL trypsin inhibitor or you could use complete protease inhibitor cocktail (Sigma #4693116001) and 1 mM PMSF.
4. 10% n-Dodecyl β-maltoside, sol-grade (DDM) and 2% cholesteryl hemisuccinate (CHS) (w/v) solution is prepared by adding 30 mL of dH_2O and 10 mL of 1 M Tris–HCl, pH 8.0 in a

50 mL falcon tube to reach a final Tris concentration of 200 mM. Weigh 5 g of DDM to the falcon tube and leave the tube on the rotator until the detergent goes into the solution. Add 1 g of CHS to the detergent solution. To solubilize the detergents, sonicate the mixture of DDM and CHS for 15–30 min at 100% amplitude with a constant pulse until the detergent solution becomes transparent. Bring the volume to 50 mL with water. The detergent solution can be stored at −20 °C until use. Every time when the stock is thawed, it should be vortexed thoroughly before use.

5. Anti-GFP nanobody-coupled Sepharose resin.
6. Glass gravity columns (Bio-Rad Econo-Columns, 1 × 20 cm).
7. AmiconUltra concentrators, 100 kDa cutoff.
8. Superose 6 Increase 10/300 GL column (Cytiva #29-0915-96).
9. Ultracentrifuge (i.e., Beckman Coulter ultracentrifuges with a Ti45 rotor).

2.6 Nanodisc Reconstitution and Cryo-EM Grid Preparation

1. Bio-beads (Bio-Rad #1523920).
2. 1-palmitoyl-2-oleoyl-glycero-3-phosphocholine (POPC, Avanti #850457).
3. A water bath sonicator.
4. Buffer A: 25 mM Tris-HCl, pH 8.0, 150 mM NaCl.
5. 10% n-Dodecyl-β-D-Maltopyranoside (DDM).
6. Quantifoil R1.2/1.3200-mesh holy carbon grids.
7. PELCO easiGlow glow discharge cleaning system.
8. Vitrobot Mark IV (FEI).

3 Methods

3.1 Protein Expression: Cx43

1. Grow HEK293F cells in Dulbecco's Modified Eagle's medium (DMEM) supplemented with 10% FBS and penicillin/streptomycin (PenStrep), in 15 cm cell culture plates at 37 °C and 5% CO_2.
2. Split the cells to the desired quantity by washing the plate with 10 mL of PBS and incubating the cells for approximately 3 min in 1.5 mL of trypsin-EDTA per plate, resuspended in appropriate volume of DMEM supplemented with 10% FBS and PenStrep and used for inoculation of multiple plates. With cells split 1:5, 80–90% confluence can be reached within 2–3 days (*see* **Note 5**).

3. Transfected when 80–90% confluent. Immediately before transfection, exchange the medium to DMEM supplemented with 2% FBS and PenStrep.
4. For PEI transfection, prepare a DNA dilution and a PEI dilution in DMEM. Volume of DNA, equivalent to 50 μg DNA per plate, is diluted in a final volume of 500 μL of DMEM per plate. Prepare the PEI solution adding 100 μL of 1 mg/mL PEI into 500 μL of final volume of DMEM (100 μg of PEI; DNA: PEI = 1:2 (w/w)) (*see* **Note 6**). The PEI and DMEM dilutions are mixed together gently, and the mixture is incubated for 5 min for formation of DNA:PEI complexes. Add the transfection mixture (1 mL) to each plate in a drop-wise manner. Place the cells in the incubator at 37 °C and 5% CO_2 for protein expression. Up to 20 plates can be transfected at a time.
5. After 48 h of expression (*see* **Note** 7), harvest the cells using a cell scraper and transfer the cells into a 50 mL falcon tube. Centrifuge the cells at 800 × *g* and 4 °C for 15 min and discard the supernatant. Resuspend the pellet in PBS and centrifugate the cells one more time. Freeze and store the cells at −80 °C until the day of the experiment.

3.2 MSP2N2 Protein Expression and Purification

The membrane scaffold protein MSP2N2 can be purchased from Sigma (Catalog number #MSP12). Alternatively, it can be purified using a modified protocol [24], as described below.

1. Transform *E. coli* BL21[DE3] with the MSP2N2 plasmid, grow the cells on the LB plates with 50 μg/mL kanamycin overnight at 37 °C.
2. The next day, inoculate a single colony into 100 mL LB medium and grow overnight at 37 °C.
3. The next day, inoculate 1 L of TB medium with 30 mL of the overnight culture, grow at 37 °C to OD_{600} of ~2.5–3.0.
4. Induce expression using 1 mM IPTG for 3 h at 37 °C. Harvest the cells by centrifugation and store the pellet at −80 °C for further purification.
5. For purification, resuspend the cell pellet in a final volume of 50 mL of lysis buffer with 1 mM PMSF and 10 μg/mL DNAse (per 1 L of cell culture).
6. Sonicate the cell suspension (Vibra-Cell Sonicator 50% amplitude, 1 s pulse) on ice for 15 min.
7. Clear the lysate by centrifugation for 30 min at 20,000 × *g*.
8. Add the cleared lysate to Ni-NTA resin (2 mL), incubate 30 min with rotation. Collect the resin into a gravity column.
9. Wash the Ni-NTA resin with 40 mL of wash buffer 1.
10. Wash the resin again with 40 mL of wash buffer 2.

11. Elute the protein with elution buffer.
12. Desalted the eluted protein into desalting buffer using a G-25 PD-10 desalting column.
13. Concentrate the protein, split the protein into multiple aliquots in Eppendorf tubes and flash-freeze the tubes in liquid nitrogen, and store them at −80 °C for further experiment. Determine the protein concentration by measuring A280; molar extinction coefficient 39,880 M^{-1} cm^{-1}; Abs 0.1% (1 g/L) = 0.876).

3.3 GFP Nanobody Expression and Purification

1. Transform *E. coli* BL21[DE3] with anti-GFP nanobody plasmid [23] and grow the cells on the LB plates with 100 μg/mL ampicillin overnight at 37 °C.
2. Inoculate 100 mL of complete TB medium with a single colony and place the flask in a shaking incubator at 37 °C and 180 rpm overnight for preculture generation (*see* **Note 2**).
3. Add 30 mL of preculture to 800 mL of complete TB medium (*see* **Note 2**) and grow cells at 37 °C and 180 rpm (*see* **Note 8**).
4. Once OD_{600} of 0.8 is reached, reduce the bacterial shaker temperature to 20 °C. Induce the protein expression using 0.5 mM IPTG. Allow the protein to express overnight (~16 h).
5. Harvest the bacterial culture by centrifugation at 4500 rpm (rotor F8S-6x1000Y, PTT) for 20 min at 4 °C. Discard the supernatant and store the pellet in a 50 mL falcon tube at −80 °C.
6. For purification, resuspend the cells in 25 mL of resuspension buffer per 1 L of bacterial pellet.
7. Lyse the cells using a microfluidizer at 18,000 psi with 3 cycles (*see* **Notes 3** and **9**).
8. Centrifuge the lysate at 19,000 rpm (rotor A8.24, Herolab) at 4 °C for 45 min to remove cell debris.
9. Collect the supernatant and incubate it with 2 mL of pre-equilibrated Ni-NTA resin per 1 L of bacterial culture for 30 min at 4 °C at constant rotation. The Ni-NTA resin is pre-equilibrated in the resuspension buffer (*see* **Note 10**).
10. Collect the Ni-NTA resin in a 50 mL gravity column and wash the resin with four column volumes of wash buffer 1, followed by a wash with four column volumes of wash buffer 2.
11. Elute the protein using the 1.5 column volumes of the elution buffer, by adding the buffer to the resin, incubating the resin for 5 min in the buffer, and collecting the eluate in an AmiconUltra concentrator with a 10 kDa filter cut-off (*see* **Note 11**).
12. Purify the protein by size-exclusion chromatography (SEC) using Superdex 75 16/600 GL column, pre-equilibrated in

SEC buffer. For SEC, inject 5 mL of concentrated protein at a time (*see* **Note 12**). After SEC, the protein is collected into a 15 mL falcon tube, to which glycerol is added to a final concentration of 10% (v/v). Measure the protein concentration (molar extinction coefficient 26,930 M^{-1} cm^{-1}; Abs 0.1% (= 1 g/L) 1.952), flash-freeze the protein in aliquots in liquid nitrogen, and store the anti-GFB nanobody at −80 °C until the day of affinity resin preparation.

3.4 Anti-GFP Nanobody Affinity Resin Preparation

1. Weigh 1 g of CNBr-Sepharose 4B beads in a 50 mL falcon tube. Add 10 mL of 1 mM HCl and incubate the beads at 4 °C with constant rotation for 30 min.
2. Collect the resin in a 50 mL gravity column.
3. Wash the resin with 20 mL of 1 mM HCl.
4. Wash the resin with 20 mL of 0.1 M $NaHCO_3$ pH 8.5, 0.5 M NaCl. Resuspend the beads in 15 mL of the same buffer and transfer them to a fresh 50 mL falcon tube.
5. Add 5–10 mg of anti-GFP nanobody per mL of swollen resin in a falcon tube. Incubate the tube on the roller-mixer at 4 °C overnight.
6. Collect the resin in a 50 mL gravity column.
7. Wash the resin twice with 20 mL of 0.2 M glycine-NaOH pH 8.0.
8. Wash the resin with 20 mL of buffer A, and after the wash, resuspend it in 10 mL buffer A, which should be transferred in a 50 mL falcon tube and stored at 4 °C (*see* **Note 13**).

3.5 Protein Purification: Cx43

1. Thaw and resuspend the cell pellets in buffer A supplemented with protease inhibitors. For a cell pellet prepared from 100 × 15 cm cell culture plates of HEK293F cells expressing Cx43-3C-YFP-twinStrep construct, use 50 mL of buffer A.
2. Lyse the cell suspensions with a Vibra-Cell Sonicator using a sonication cycle of 0.5 s pulse with a 0.5 s break at 35% amplitude for 4 min.
3. The cell membranes are collected by ultracentrifugation at 35,000 rpm (Ti45 rotor, Beckman Coulter) for 50 min at 4 °C.
4. Discard the supernatant and resuspend the pellet in buffer A supplemented with protease inhibitors using an Ultra Turrax homogenizer.
5. Add DDM-CHS stock solution to a final concentration of 1% DDM and 0.2% CHS (*see* **Note 14**) to the resuspended membranes for solubilization for 1 h at 4 °C with constant rotation.

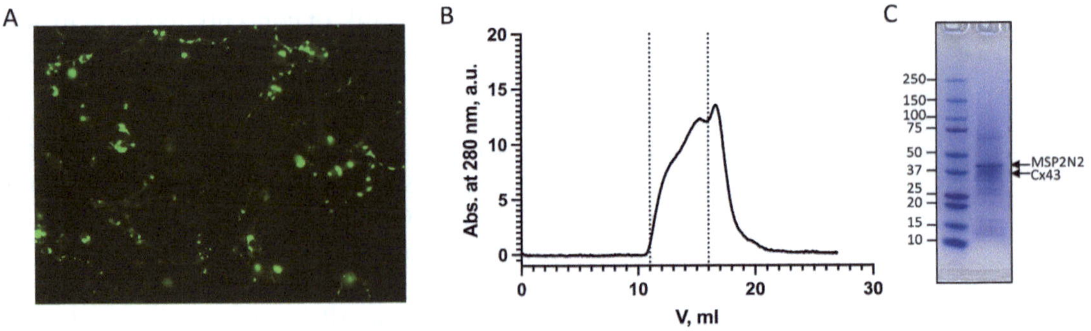

Fig. 2 Expression and purification of Cx43 in MSP2N2 nanodiscs. (**a**) The fluorescence image of Cx43 with a YFP tag expressed in HEK293 cells shows a typical expression pattern, 24 h after transfection; expression of the fluorescent protein fusion was performed using Bio-Rad ZOE Fluorescent Cell Imager. (**b**) The size-exclusion chromatography (SEC) profile of the purified Cx43 reconstituted into MSP2N2 nanodiscs; the dotted lines indicate the broad SEC peak taken for further experiments (which includes both Cx43 GJCs and HCs). (**c**) SDS-PAGE of Cx43 reconstituted in MSP2N2 nanodiscs. The arrows indicate the positions of MSP2N2 and Cx43; the numbers correspond to the molecular weight (MW, kDa), based on the left lane (MW marker)

6. Clear the lysates by ultracentrifugation at 35,000 rpm (Ti45 rotor, Beckman Coulter) for 50 min at 4 °C and collect the supernatant.
7. Add 2 mL of anti-GFP nanobody-coupled Sepharose to the lysate (prepared from 100 plates) and incubate for 30 min at 4 °C with constant rotation.
8. Collect the resin into a gravity column (*see* **Note 15**).
9. Wash the resin using 40 column volumes of buffer B. The wash should be performed in four steps of 10 column volumes each with a slow flow to allow efficient detergent exchange.
10. To elute the protein, add 10 mL of buffer B to the closed gravity column and add HRV 3C protease (200 mg) for cleavage. Seal the column tightly and leave it to rotate slowly overnight at 4 °C. From this step, proceed directly to "3.6 Nanodisc reconstitution." Alternatively, **steps 11–13** describe the procedure for completing the purification of the Cx43 sample in detergent (digitonin).
11. Collect the solution containing the eluted protein into an Amicon Ultra concentrator with a 100 kDa filter cut-off and concentrate the protein by centrifugation at 1000 × *g* and 4 °C, using 2 min spins at a time. In between each spin, the concentrator should be inverted 4–5 times gently to mix the concentrated protein. Concentrate the protein to a volume corresponding to that needed to fully fill the HPLC injection loop (*see* **Note 16**).

12. Prior to SEC centrifuge, the protein in a microcentrifuge at 20,000 × *g* and 4 °C for 5 min to remove any precipitated detergent or protein.
13. Purify the sample further by SEC by injecting 1 mL of the concentrated protein onto a Superose 6 Increase 10/300 GL column pre-equilibrated with buffer B (*see* **Note 17**).
14. Collect the SEC peak fractions that correspond to the purified Cx43 (Fig. 2). Measure protein concentration by measuring the protein absorbance at 280 nm (molar extinction coefficient 56,840 M^{-1} cm^{-1}; Abs 0.1% (= 1 g/L) 1.333).

3.6 Nanodisc Reconstitution

1. Prepare POPC for nanodisc reconstitution: (i) dissolve the lipid in chloroform if the lipid is powder. Use nitrogen gas to dry the lipid in a glass tube; (ii) rotate the tube during drying to let the lipid distribute uniformly on the surface of the glass tube; (iii) leave the glass tube in the lyophilizer overnight to totally dry the lipid; (iv) resuspend the lipid in buffer 25 mM Tris-HCl pH 8.0, 150 mM NaCl, 1% DDM; the final lipid concentration should be 10 mg/mL; (v) sonicate the lipid in a water bath sonicator until the lipid becomes transparent (*see* **Note 18**).
2. Thaw an aliquot of MSP2N2 protein (frozen protein can be kept in a −80 °C freezer until the day of experiment), and keep it on ice.
3. Concentrate Cx43 collected after GFP nanobody purification (Subheading 3.5, **step 10**) to ~300 μL.
4. Add lipids and the membrane scaffold protein (MSP2N2) to the concentrated Cx43 at a ratio of 6:4:300 (Cx43:MSP2N2: POPC). The sequence is as follows: mix POPC with Cx43 first and incubate for 30 min at room temperature. Then add MSP2N2 to the mixture and repeat the incubation step.
5. Add Bio-beads (0.5 mg) to the reconstitution system. The Bio-beads must be washed with methanol and buffer A prior to addition to the protein (see **Notes 19** and **20**).
6. Move the sample to 4 °C and rotate overnight to remove the detergent.
7. Collect the supernatant by gentle pipetting, keeping the Bio-beads out of the protein/lipid mixture.
8. Clear the protein/lipid mixture by ultracentrifugation at 100,000 × *g* and 4 °C for 30 min.
9. Inject 1 mL of the mixture onto a Superose 6 Increase 10/300 GL column pre-equilibrated with buffer A. Collect the peak fractions corresponding to Cx43 in MSP2N2 nanodiscs.
10. Determine the concentration of the protein preparation, using the extinction coefficient 1.3. The purified and reconstituted

protein can be used for biochemical or biophysical experiments, or for cryo-EM grid preparation. The protein, concentrated to 2 mg/mL for cryo-EM grid preparation using a 100 kDa cut-off concentrator, can be used for cryo-EM analysis.

3.7 Cryo-EM Grid Preparation and Data Collection

1. Glow discharge the Quantifoil R1.2/1.3200-mesh grid at 25 mA for 30 s using PELCO easiGlow glow discharge device.
2. Cool the Vitrobot Mark IV to 4 °C and set it to 100% humidity. Leave the filter paper in the Vitrobot for at least 20 min. Prepare the liquid ethane.
3. Apply 3.5 μL of the purified Cx43 nanodisc sample to the carbon side of the grid. Blot the grid with force 0 and blot time 3 s, without any waiting time or drain time, and immediately plunge the grid into the liquid ethane.
4. Transfer the grid under cryogenic conditions into liquid nitrogen for further storage or cryo-EM data collection.

4 Notes

1. In case cells other than HEK293F are used for protein expression, various parameters may need to be optimized, including: DNA concentration and DNA:PEI ratio, expression medium, expression time, and temperature.
2. TB medium can be replaced with other types of media for bacterial growth (e.g., LB).
3. French press, sonication, or other techniques can be used as alternative methods for cell disruption.
4. The amount of 1 g of CNBr sepharose beads corresponds to ~3.5 mL of swollen medium.
5. Depending on the cell line, the medium, and other cell culture conditions, the time necessary to reach the appropriate confluency may vary. Optimal parameters have to be established empirically.
6. The DNA concentration used for transfection can be optimized around 50 μg of DNA per plate depending on cell line, vector, etc. The DNA to PEI ratio can also be optimized, but a ratio of 1:2 (w/w) produces consistently satisfactory results.
7. The expression time as well as the optimal expression temperature can vary depending on the vector and cell line used. For optimization of expression time, a window of 24–72 h is recommended.
8. Larger biomass can be generated using TB medium, compared to the LB medium.

9. The cells need to be completely resuspended to not clog the microfluidizer, e.g., using a homogenizer.
10. Fresh Ni-NTA resin produces the most consistent results. The resin regenerated following the vendor's instructions can be reused multiple times; reduction of protein yield may be an indication of poor resin quality (e.g., due to poor resin regeneration).
11. The eluate can also be collected in a 50 mL falcon; glycerol can be added to a final concentration of 10% (v/v), and the sample can be flash frozen and stored at −80 °C. Aliquot the protein and avoid multiple freeze-thaw cycles whenever possible. The protein should not be frozen at a concentration above 20 mg/mL to avoid protein precipitation upon thawing.
12. Columns other than Superdex 75 10/100 GL can be used. The choice of the column depends on the ability to resolve the SEC peak that corresponds to the nanobody.
13. Storage times longer than 1 week are not recommended. Longer storage time may lead to deterioration of the affinity resin, resulting in poor protein yield.
14. If a different connexin isoform is purified, the detergent type and concentration need to be optimized for each of the purification steps (solubilization, wash, elution, and SEC).
15. The gravity column should be preferably small to prevent excessive mixing and frothing of the solution, which may lead to protein aggregation and reduction of the final yield.
16. When purifying Cx43 from 100 plates, concentrating the protein to a volume of 500 μL–1 mL produces good results. Concentrating the protein to lower volumes is not necessary as this step.
17. SEC of the protein can also be executed on other equivalent columns. The injection volume depends on the available injection loops.
18. Be careful when you dry the lipid. Perform the experiment in the fume hood and avoid the lipid spillage from the glass tube. It will take time to sonicate the lipid until transparency, depending on the power of the sonicator.
19. Chloroform and methanol are toxic organic solvents. Be cautious, dispose of these reagents into the appropriate organic waste receptacle.
20. We use an excess of Bio-beads to make sure we get rid of the detergent. Based on the manufacturer's instructions, 1 g of Bio-beads can adsorb 0.07 × g of TritonX-100 (a standard detergent).

References

1. Rodriguez-Sinovas A, Sanchez JA, Valls-Lacalle L, Consegal M, Ferreira-Gonzalez I (2021) Connexins in the heart: regulation, function and involvement in cardiac disease. Int J Mol Sci 22(9). https://doi.org/10.3390/ijms22094413
2. Robertson JD (1963) The occurrence of a subunit pattern in the unit membranes of Club endings in Mauthner cell synapses in goldfish brains. J Cell Biol 19:201–221. https://doi.org/10.1083/jcb.19.1.201
3. Laird DW, Lampe PD (2018) Therapeutic strategies targeting connexins. Nat Rev Drug Discov 17(12):905–921. https://doi.org/10.1038/nrd.2018.138
4. Beyer EC, Paul DL, Goodenough DA (1987) Connexin43: a protein from rat heart homologous to a gap junction protein from liver. J Cell Biol 105(6 Pt 1):2621–2629. https://doi.org/10.1083/jcb.105.6.2621
5. Poelzing S, Rosenbaum DS (2004) Altered connexin43 expression produces arrhythmia substrate in heart failure. Am J Physiol Heart Circ Physiol 287(4):H1762–H1770. https://doi.org/10.1152/ajpheart.00346.2004
6. Liang Z, Wang X, Hao Y, Qiu L, Lou Y, Zhang Y et al (2020) The multifaceted role of astrocyte Connexin 43 in ischemic stroke through forming Hemichannels and gap junctions. Front Neurol 11:703. https://doi.org/10.3389/fneur.2020.00703
7. D'Hondt C, Iyyathurai J, Himpens B, Leybaert L, Bultynck G (2014) Cx43-hemichannel function and regulation in physiology and pathophysiology: insights from the bovine corneal endothelial cell system and beyond. Front Physiol 5:348. https://doi.org/10.3389/fphys.2014.00348
8. Xing L, Yang T, Cui S, Chen G (2019) Connexin Hemichannels in astrocytes: role in CNS disorders. Front Mol Neurosci 12:23. https://doi.org/10.3389/fnmol.2019.00023
9. Paznekas WA, Boyadjiev SA, Shapiro RE, Daniels O, Wollnik B, Keegan CE et al (2003) Connexin 43 (GJA1) mutations cause the pleiotropic phenotype of oculodentodigital dysplasia. Am J Hum Genet 72(2):408–418. https://doi.org/10.1086/346090
10. Paznekas WA, Karczeski B, Vermeer S, Lowry RB, Delatycki M, Laurence F et al (2009) GJA1 mutations, variants, and connexin 43 dysfunction as it relates to the oculodentodigital dysplasia phenotype. Hum Mutat 30(5): 724–733. https://doi.org/10.1002/humu.20958
11. Wiest T, Herrmann O, Stogbauer F, Grasshoff U, Enders H, Koch MJ et al (2006) Clinical and genetic variability of oculodentodigital dysplasia. Clin Genet 70(1):71–72. https://doi.org/10.1111/j.1399-0004.2006.00631.x
12. Pizzuti A, Flex E, Mingarelli R, Salpietro C, Zelante L, Dallapiccola B (2004) A homozygous GJA1 gene mutation causes a Hallermann-Streiff/ODDD spectrum phenotype. Hum Mutat 23(3):286. https://doi.org/10.1002/humu.9220
13. Dasgupta C, Martinez AM, Zuppan CW, Shah MM, Bailey LL, Fletcher WH (2001) Identification of connexin43 (alpha1) gap junction gene mutations in patients with hypoplastic left heart syndrome by denaturing gradient gel electrophoresis (DGGE). Mutat Res 479(1–2):173–186. https://doi.org/10.1016/s0027-5107(01)00160-9
14. Unger VM, Kumar NM, Gilula NB, Yeager M (1999) Three-dimensional structure of a recombinant gap junction membrane channel. Science 283(5405):1176–1180. https://doi.org/10.1126/science.283.5405.1176
15. Maeda S, Nakagawa S, Suga M, Yamashita E, Oshima A, Fujiyoshi Y et al (2009) Structure of the connexin 26 gap junction channel at 3.5 A resolution. Nature 458(7238):597–602. https://doi.org/10.1038/nature07869
16. Myers JB, Haddad BG, O'Neill SE, Chorev DS, Yoshioka CC, Robinson CV et al (2018) Structure of native lens connexin 46/50 intercellular channels by cryo-EM. Nature 564(7736):372–377. https://doi.org/10.1038/s41586-018-0786-7
17. Flores JA, Haddad BG, Dolan KA, Myers JB, Yoshioka CC, Copperman J et al (2020) Connexin-46/50 in a dynamic lipid environment resolved by CryoEM at 1.9 A. Nat Commun 11(1):4331. https://doi.org/10.1038/s41467-020-18120-5
18. Lee HJ, Jeong H, Hyun J, Ryu B, Park K, Lim HH et al (2020) Cryo-EM structure of human Cx31.3/GJC3 connexin hemichannel. Sci Adv 6(35):eaba4996. https://doi.org/10.1126/sciadv.aba4996
19. Khan AK, Jagielnicki M, Bennett BC, Purdy MD, Yeager M (2021) Cryo-EM structure of an open conformation of a gap junction hemichannel in lipid bilayer nanodiscs. Structure 29(9):1040–7 e3. https://doi.org/10.1016/j.str.2021.05.010
20. Qi C, Gutierrez SA, Lavriha P, Othman A, Pigozzi DL, Bayraktar E et al (2023) Structure

of the connexin-43 gap junction channel in a putative closed state. Elife 12:RP87616. https://doi.org/10.7554/eLife.87616.3

21. Denisov IG, Sligar SG (2016) Nanodiscs for structural and functional studies of membrane proteins. Nat Struct Mol Biol 23(6):481–486. https://doi.org/10.1038/nsmb.3195
22. Grinkova YV, Denisov IG, Sligar SG (2010) Engineering extended membrane scaffold proteins for self-assembly of soluble nanoscale lipid bilayers. Protein Eng Des Sel 23(11):843–848. https://doi.org/10.1093/protein/gzq060
23. Kubala MH, Kovtun O, Alexandrov K, Collins BM (2010) Structural and thermodynamic analysis of the GFP:GFP-nanobody complex. Protein Sci 19(12):2389–2401. https://doi.org/10.1002/pro.519
24. Ritchie TK, Grinkova YV, Bayburt TH, Denisov IG, Zolnerciks JK, Atkins WM et al (2009) Chapter 11—reconstitution of membrane proteins in phospholipid bilayer nanodiscs. Methods Enzymol 464:211–231. https://doi.org/10.1016/S0076-6879(09)64011-8

Chapter 4

Molecular Dynamics Simulation of Permeation Through Connexin Channels

Francesco Zonta, Fabio Mammano, and Sergio Pantano

Abstract

Molecular dynamics (MD) simulations are a collection of computational tools that can be used to trace intermolecular interactions at the sub-nanometer level. They offer possibilities that are often unavailable to experimental methods, making MD an ideal complementary technique for the understanding a plethora of biological processes. Thanks to significant efforts by many groups of developers around the world, setting up and running MD simulations has become progressively simpler. However, simulating ionic permeation through membrane channels still presents significant caveats.

MD simulations of connexin (Cx) hemichannels (HCs) are particularly problematic because HCs create wide pores in the plasma membrane, and the lateral sizes of the extracellular and intracellular regions are quite different. In this chapter, we provide a detailed instruction to perform MD simulations aimed at computationally modeling the permeation of inorganic ions and larger molecules through Cx HCs.

Key words Connexon, Free energy barriers, Permeation, Conduction, Pore

1 Introduction

Connexins (Cxs) are integral membrane proteins composed of four transmembrane (TM) helixes connected by two extracellular loops (EC1, EC2) and one cytoplasmic loop (CL) (Fig. 1). Cx hemichannels (HCs), also known as connexons, are hexameric arrays of Cx proteins. An HC can operate as a large-pore plasma membrane channel [1–4] or dock head-to-head in the extracellular space with another HC to form an intercellular gap junction channel (IGJCs) [5–8]. IGJCs cluster in crowded plaques containing thousands of copies and connect the cytoplasmic spaces of two neighboring cells [9, 10] (Fig. 1). For almost a decade, the only known Cx structure was that of the Cx26 IGJC, but recently other isoforms have been reported [11–18].

HCs and IGJCs are responsible for exchanging a plethora of molecular signals, ranging from single ions to small RNA [10],

Fabio Mammano and Mauricio Retamal (eds.), *Connexin Hemichannels: Methods and Protocols*, Methods in Molecular Biology, vol. 2801, https://doi.org/10.1007/978-1-0716-3842-2_4,

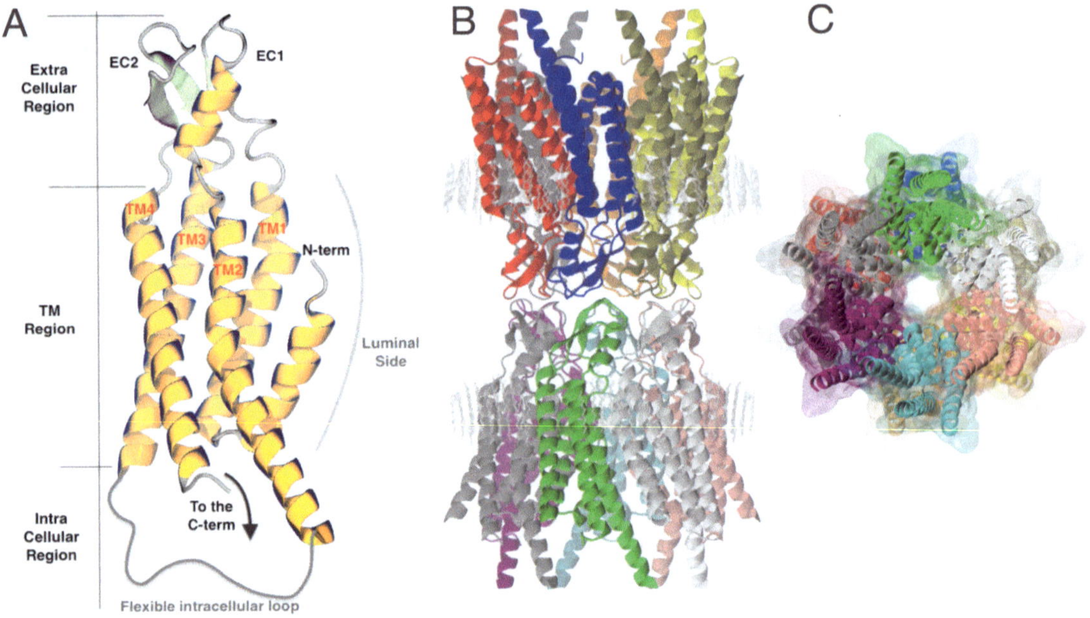

Fig. 1 Topology and assembly of connexin proteins

between cytosol and extracellular milieu (HCs) or between the cytosols of abutting cells (IGJCs). Due to their crucial involvement in virtually all aspects of cell-to-cell communication, HCs and IGJCs are implicated in variety of physiological as well as pathological processes. The latter include cancer, inflammation, neurodegenerative diseases, hereditary diseases, etc. [17].

Despite significant development in MD and the effort made by several groups to facilitate the setting up and running of numerical simulations [18, 19], simulating permeation processes through proteinaceous pores remains an intricate problem with many caveats. The aim of this chapter is to ease the task for Cx HCs.

2 Materials

2.1 3D Structures and Software

Several MD software packages, which make MD simulations easily accessible even to nonexperts, are currently available. Amber [20], Gromacs [21], CHARMM [22], and OpenMM [23] are probably the most widely used (*see* **Note 1**).

More important is the choice of the force field. From our experience, we suggest using the Amber ff19SB [24] or the CHARMM36 force field [25] for atomistic simulations and the SIRAH force field for coarse-grained simulations [26, 27].

To begin with, the coordinates of a given Cx channel can be retrieved from the Protein Data Bank (PDB, https://www.rcsb.pdb). Currently, the structures of human or sheep Cx26, Cx31.3,

Cx32, Cx43, Cx46, and Cx50 HCs at different resolutions have been deposited on the PDB [11, 12, 14, 15, 100, 200]. None of these structures is complete as they all contain missing regions, especially CL and the (unstructured) CT. Hence, these missing domains must be fitted in before starting the simulation (*see* **Note 2**). Any of the methods reported in **Note 1** will produce energetically reasonable conformations for the CL. However, the situation is different for the CT, as all Cxs except Cx26 have a long, intrinsically disordered CT that is supposed to become partially structured when interacting with other molecules (e.g., Calmodulin). While protein structure prediction has improved dramatically in the last years (to the point that the protein folding problem can be considered technically solved [30–32]), modeling and understanding the dynamical properties of long, unstructured protein domains remains an elusive problem.

The protonation state of side chains must also be carefully adjusted to the desired pH (*see* **Note 3**). Connexins also contain disulfide bridges among highly conserved cysteines. Although some of these features can be automatically recognized and fixed by the module pdb4amber, the user must ensure that every detail is considered before starting a MD simulation (*see* **Note 4**).

Finally, connexin proteins undergo posttranslational modifications (PTMs) [31, 32] which should be taken into account as they may affect the permeation process under study [35–37].

3 Methods

3.1 Inserting HCs in the Plasma Membrane

Once the 3D model has been completed, we suggest running an energy minimization in vacuo to relax possible interatomic clashes (*see* **Note 5**). Next, the HC should be inserted into a patch of phospholipids mimicking the plasma membrane (*see* **Note 6** and Fig. 2). This is slightly more complicated for Cx HCs than for most other membrane proteins, owing to the difference in diameter between the extracellular and intracellular sides of the channel.

When running simulations in periodic boundary conditions, this size difference results in a different number of phospholipids on the two sides of the membrane, producing pressure imbalances if the system was not equilibrated properly [36] (*see* **Note** 7). The final preparatory step is to add water and ions to mimic a physiological ionic strength (*see* **Note 8**). Once this has been accomplished, the user must perform an initial equilibration of the molecular system (*see* **Note 9**) which includes:

1. Energy Minimization (*see* **Note 10**).
2. First equilibration MD (on the order of 1 ns) at constant volume (NTV ensemble, *see* **Note 11**).

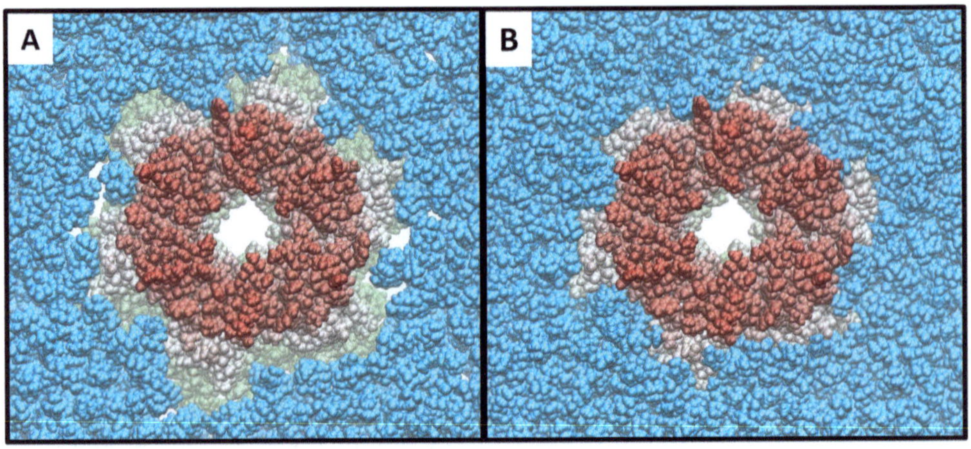

Fig. 2 Relaxation of the membrane after the connexon insertion

3. Second equilibration MD simulation at constant pressure (NPT ensemble) with anisotropic barostat, i.e., the pressure components along the axis of the channel (Z) and orthogonal to the axis of the channel (X, Y) are coupled to independent pistons of a barostat (*see* **Note 12**).

See also **Note 13**. It is important to check that the cavity is effectively full of solvent before removing positional restraints from the protein. If not, the lateral pressure could squeeze the channel producing spurious distortions. If the cavity is not full of water molecules and ions, additional equilibration simulations should be carried out.

3.2 Equilibrium MD in the NPT Ensemble

Successful completion of the equilibration stage described in the previous section sets the ground for the production runs under standard conditions. There are no particular requirements or further setup steps specific for a Cx HC system compared to other membrane proteins. As explained earlier, we recommend running simulations in the NPT ensemble with an anisotropic pressure couple.

The only caveat regards the choice of ions in the solvent. When using standard Periodic Boundary Conditions in a membrane system, the intercellular and extracellular media are in direct contact, so it is impossible to maintain different ion concentrations between the two sides, as it happens in physiological situations. Workarounds for these issues include adding a second membrane, a vacuum layer, or an external potential that limits ions mobility along the Z-axis [36] (for IGJCs, the problem is automatically solved by the presence of a second membrane). The other issue in the choice of the solvent concerns the presence of calcium ions (Ca^{2+}). It is well known that Ca^{2+} is essential for the regulation and gating of HCs and IGJCs. However, the finite number of water molecules required to solvate a system of the size of the HCs implies that even a few (<10) calcium ions would produce a very high concentration in typical simulation boxes (*see* **Note 14**). This could generate important artifacts in the simulations, as Ca^{2+} could stick for relatively long time periods to the side chains of acidic amino acids and to phosphate groups in phospholipids. These effects have to be carefully considered when analyzing the results.

Finally, the system must be adapted to the particular problem under investigation. For example, the membrane composition should be adjusted if one is interested in analyzing the interplay between the Cx proteins and specific phospholipids or cholesterol.

Nowadays, MD simulations in the order of the microsecond can be conducted to achieve a reasonable sampling size of processes associated with water and ionic permeation through Connexin channels (*see* **Note 15**).

3.3 Permeability Calculations and Enhanced Sampling

A direct comparison of computed ionic and molecular permeability through the channel would require the capacity of simulating time windows comparable with the resolution of patch-clamp experiments, which is in the order of the milliseconds. MD simulations of such duration are at the very edge of current computational capabilities. This problem, which is share by every membrane channel, is exacerbated by the size of HC pores and by the possible existence of conformational transitions which have not been structurally well characterized yet [37].

In the following, we describe several methods that have been applied in the literature to produce estimates of conductance and permeability, and their limitations.

The first method is a direct estimate of conductance. In principle, it is possible to obtain such an estimate by counting the number of ions passing through the channel in a given amount of time, under the effect of a given external potential. Such method has been recently applied to evaluate permeation properties of a Cx31.3 hemichannel [14], and require microseconds long simulation to get reliable average currents (*see* **Note 15**).

Longer timescales can be easily reached using coarse-grained force fields. Using the SIRAH force field [38], simulations of tens of microseconds showed that ion permeation does not happen in an ordered or regular fashion. Instead, ions traverse the pore in bursts resembling volcano eruptions, followed by low activity time intervals in which ions accumulate in the pore vestibule. The electrostatic repulsion then works as an accumulated pressure that is released transiently during a burst [37]. However, coarse-grained force fields may find difficulties to correctly reproduce kinetic properties of the system, since the heights of energy barriers between different states may not be comparable to those of atomistic potentials.

Besides these direct approaches, earlier works showed a certain level of success in predicting ratios between the permeability of different molecules or the permeability of the same species across

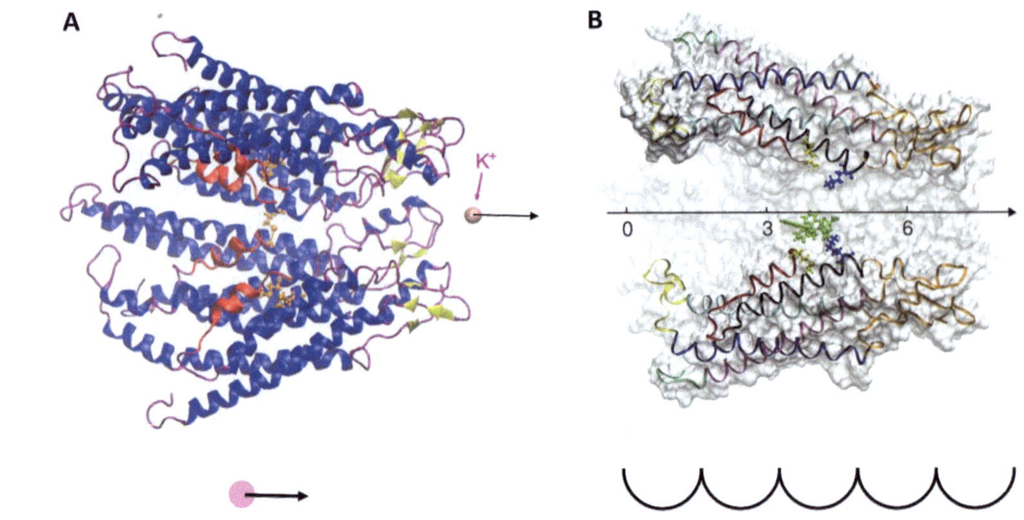

Fig. 3 Ion pulling and umbrella sampling of molecules

different connexin hemichannels using various enhanced sampling strategies [33, 39, 40]. Such simulations are less expensive from a computational point of view but still reliable because averaging performed over independent simulations favors error cancellation. Moreover, predictions on energy barriers depend on the calculation of free energy (or Potential of mean force) differences and not their absolute value [41].

Free energy profiles can be obtained from long unbiased trajectories by computing the probability that a given ion or molecule is found in particular positions or by using biased simulations that force the ion or molecule to move along the desired direction (the reaction coordinate, which in our case coincides with the axis of the channel). For example, by slowly pulling potassium ions (K^+) along the axis of the channel, it is possible to compute the mean force that the ions experience and calculate the free energy from it (*see* **Note 16**).

Computing the free energy of permeation for molecules is a more complicated task, as the time scale of these processes goes from hundreds of microseconds to milliseconds. They require a comprehensive sampling of the internal degrees of freedom, as each molecule could have a preferred orientation for the permeation. This can be done by using umbrella sampling methods [42, 43] (*see* Fig. 3 and **Note 17**), which force the molecule at a particular point of the *Z*-direction (collinear with the pore axis) without constraining the position on the *XY* plane and the internal degrees of freedom [34, 43].

However, even these advanced sampling methodologies may not ensure a complete sampling of the configuration space, as entropic barriers can be high because the restricted space in the narrower part of the channel may impair rotations.

3.4 Analysis of MD Trajectories and Limitations

Analysis of MD trajectories is never a trivial task, as many degrees of freedom are evolving during the dynamics, and one should have a clear working hypothesis before designing an appropriate simulation protocol. Cx systems are not special in this sense, and data analysis methods that work for any standard MD simulation can be applied.

It is a good habit to perform some "sanity checks" on the quality of the simulations, by analyzing standard observables such as the Root Mean Square Deviation and the Root Mean Square Fluctuations. Their expected behavior should be similar to that reported in Figure 3a and Figure 5 of reference [39].

Furthermore, it is a good practice to perform few replicas of a simulation to eliminate the possibility that the trajectory is drifting away from the expected behavior. This is of particular importance for out of equilibrium simulations (such as umbrella sampling simulations), because results can be heavily dependent on small fluctuations of the initial coordinates.

It is also important to keep in mind that the less structurally characterized regions of the Cx HCs (CL, CT) are targets of a number of posttransductional modifications with significant regulatory roles. Although the modeling tools for loops and C-termini mentioned above provide energetically reasonable conformers, they are reconstructed mainly to preserve the continuity of the polypeptidic chain. Therefore, these regions should be excluded from root mean square deviations/fluctuations, and special attention should be paid to avoid overinterpreting intermolecular interactions involving those regions. Moreover, HCs and IGJCs are regulated by pH, the effect of which is usually neglected in MD simulations whereby the protonation state is topologically fixed during the entire simulation time.

4 Notes

1. The various MD software packages have differences in algorithm implementation and speed. They are all actively developed, and newer versions have a tendency to perform better and be more user friendly. Nevertheless, they are all accepted by the computational biology community, and the final choice can be ascribed to personal preferences.
2. Missing regions in protein structures reported in the PDB can be reconstructed using homology protein software. Web interfaces such as ModLoop (https://modbase.compbio.ucsf.edu/modloop/), Lobo (https://protein.bio.unipd.it/services), or SwissModel (https://swissmodel.expasy.org) can be easily used to provide a reasonable and complete starting conformation of proteins. Alternatively, AI predictions made by Alphafold are provided in the UniProt site for each protein.

3. Using the AMBER package, protonation states are set to neutral pH by default (pH = 7). Moreover, histidine amino acids are protonated in epsilon position unless otherwise stated. This may be indicated by changing the residue name from HIS to HIE (protonated in epsilon position), HID (protonated in delta position), or HIP (positively charged). A good alternative is to submit your PDB structure to the server https://www.poissonboltzmann.org and setting the input and output formats to AMBER. It will return a properly protonated structural model at the pH selected in the input.
4. It is of outmost importance to check visually the protein before starting a simulation. ALWAYS visualize your starting model before building up a simulation box and check it for possible missing atoms, disulfide bridges, protonation, etc.
5. Due to the analytical expression of the Lennard–Jones potential, atoms that are too close to each other can generate huge values of the interaction potential. If not properly relaxed, interatomic clashes will create large distortions in the protein structure during the dynamics. In most cases, standard energy minimization algorithms are able to correctly relax a structure that has been reconstructed from a good resolution experimental structure. We suggest to perform 500 steps of steepest descent algorithm followed by 500 steps of conjugated gradient algorithm using your preferred MD software package.
6. Our advice is to insert the HC in a previously relaxed membrane patch at least 2 times larger than the desired size of the final system, so that there are enough phospholipid molecules to relax the lateral components of the pressure (i.e., the components orthogonal to the channel). Any phospholipid in close contact (<0.1 nm) with the protein and any phospholipid within the pore must be removed from the complex before equilibration can start. This procedure is likely to leave some space between the protein and the membrane, especially in proximity of the extracellular side of the channel. When solvated, this empty space is likely to be filled with water molecules, but they should quickly be expelled during the equilibration phase.
7. This could be done by removing phospholipids within 0.1 nm of connexin proteins to avoid clashes that are difficult to minimize. This leads to a difference in the number of phospholipids on the two sides of the membrane and could produce different densities in very small systems. This can be solved by equilibrating the system within a membrane patch larger than the desired one, and cutting it to the desired size once the membrane is well adapted around the proteins. Note that the

borders of the membrane must be equilibrated again to obtain correct periodic boundary conditions.

8. There are several alternatives to estimate the number of ions required to reproduce a physiological ionic strength. We suggest the method discussed in reference [44].
9. Although the workflow may seem intricate, it is actually not time consuming in comparison with the production stage, and it is essential to relax the system, avoiding artifacts during the simulation.
10. This is aimed to relax steric clashes. The default option in Amber (including a few steps using the Steepest Descend minimization algorithm to then change to the Conjugated gradient to improve convergence) works perfectly fine for this case.
11. The equilibration MD at constant volume must be conducted by imposing positional harmonic restraints on the location of the alpha carbon of the backbone to preserve the structure of the channel while the system relaxes. Force constants in the order of 10,000 kJ/mol/nm^2 are sufficient to the ensure conservation of the protein structure. This step is redundant if the model is well built, as the following step (NPT relaxation, *see* **Note 12**) could be enough to obtain a good starting configuration.
12. For NPT equilibration, use the same restraint protocol suggested in **Note 11**. This step ensures that the membrane will correctly fit around the protein, and water and ions will be expelled from the membrane, without altering the structure of the channel before running the production simulation. It is possible that few water molecules or ions still are trapped in the membrane at the end of this stage, but they can be manually removed before running another short equilibration simulation.
13. All these steps can be significantly simplified by the use of the CharmmGUI server (https://charmm-gui.org), which performs several quality checks in the user structure and may provide a prestabilized configuration. Anyhow, thorough controls are always needed (*see* **Note 3**).
14. Intracellular Ca^{2+} concentration is roughly in the order of 100 nM, while it its about 4 orders of magnitude higher in the extracellular milieu. If one ionic pair in 5500 water molecules roughly equals 1 M, this means that only a few calcium ions should be present in a state-of-the-art simulation box.
15. For a system of the size of a Cx HC in a membrane patch (150,000–200,000 atoms), a modern GPU-based workstation can produce tens of nanoseconds of dynamics per day, with a 2 fs timestep. Microseconds simulations are now affordable in

weeks to months of production runs, even without access to dedicated clusters.

16. The velocity of the pull should be as small as possible to ensure that the mean pulling force is well equilibrated. From our experience, an appropriate speed should not exceed 0.5 nm/ns.
17. Umbrella sampling is widely used, and most MD software packages have dedicated routines and tutorial for its implementation.

Acknowledgments

This work was supported by the Italian Ministry of Research and University (grant PRIN 20175C22WM to F.M.), Fondazione Telethon (grant GGP19148 to F.M.), FOCEM (MERCOSUR Structural Convergence Fund, grant COF 03/11 to S.P.), and Xi'an Jiaotong-Liverpool University (Research Development Fund RDF-23-01-026 to F.Z).

References

1. Bennett MV, Contreras JE, Bukauskas FF, Saez JC (2003) New roles for astrocytes: gap junction hemichannels have something to communicate. Trends Neurosci 26(11):610–617. https://doi.org/10.1016/j.tins.2003.09.008
2. Saez JC, Retamal MA, Basilio D, Bukauskas FF, Bennett MV (2005) Connexin-based gap junction hemichannels: gating mechanisms. Biochim Biophys Acta 1711(2):215–224. https://doi.org/10.1016/j.bbamem.2005.01.014
3. Evans WH, De Vuyst E, Leybaert L (2006) The gap junction cellular internet: connexin hemichannels enter the signalling limelight. Biochem J 397(1):1–14. https://doi.org/10.1042/BJ20060175
4. Saez JC, Schalper KA, Retamal MA, Orellana JA, Shoji KF, Bennett MV (2010) Cell membrane permeabilization via connexin hemichannels in living and dying cells. Exp Cell Res 316(15):2377–2389. https://doi.org/10.1016/j.yexcr.2010.05.026
5. Mammano F (2019) Inner ear Connexin channels: roles in development and maintenance of Cochlear function. Cold Spring Harb Perspect Med 9(7). https://doi.org/10.1101/cshperspect.a033233
6. Perkins GA, Goodenough DA, Sosinsky GE (1998) Formation of the gap junction intercellular channel requires a 30 degree rotation for interdigitating two apposing connexons. J Mol Biol 277(2):171–177. https://doi.org/10.1006/jmbi.1997.1580
7. Unger VM, Kumar NM, Gilula NB, Yeager M (1999) Three-dimensional structure of a recombinant gap junction membrane channel. Science 283(5405):1176–1180. https://doi.org/10.1126/science.283.5405.1176
8. Kumar NM, Gilula NB (1996) The gap junction communication channel. Cell 84(3):381–388. https://doi.org/10.1016/s0092-8674(00)81282-9
9. Goodenough DA, Paul DL (2009) Gap junctions. Cold Spring Harb Perspect Biol 1(1):a002576. https://doi.org/10.1101/cshperspect.a002576
10. Esseltine JL, Laird DW (2016) Next-generation Connexin and Pannexin cell biology. Trends Cell Biol 26(12):944–955. https://doi.org/10.1016/j.tcb.2016.06.003
11. Maeda S, Nakagawa S, Suga M, Yamashita E, Oshima A, Fujiyoshi Y, Tsukihara T (2009) Structure of the connexin 26 gap junction channel at 3.5 A resolution. Nature 458(7238):597–602. https://doi.org/10.1038/nature07869
12. Bennett BC, Purdy MD, Baker KA, Acharya C, McIntire WE, Stevens RC, Zhang Q, Harris AL, Abagyan R, Yeager M (2016) An electrostatic mechanism for Ca(2+)-mediated regulation of gap junction channels. Nat Commun 7:

8770. https://doi.org/10.1038/ncomms9770
13. Myers JB, Haddad BG, O'Neill SE, Chorev DS, Yoshioka CC, Robinson CV, Zuckerman DM, Reichow SL (2018) Structure of native lens connexin 46/50 intercellular channels by cryo-EM. Nature 564(7736):372–377. https://doi.org/10.1038/s41586-018-0786-7
14. Lee HJ, Jeong H, Hyun J, Ryu B, Park K, Lim HH, Yoo J, Woo JS (2020) Cryo-EM structure of human Cx31.3/GJC3 connexin hemichannel. Sci Adv 6(35):eaba4996. https://doi.org/10.1126/sciadv.aba4996
15. Flores JA, Haddad BG, Dolan KA, Myers JB, Yoshioka CC, Copperman J, Zuckerman DM, Reichow SL (2020) Connexin-46/50 in a dynamic lipid environment resolved by CryoEM at 1.9 A. Nat Commun 11(1):4331. https://doi.org/10.1038/s41467-020-18120-5
16. Brotherton DH, Savva CG, Ragan TJ, Dale N, Cameron AD (2022) Conformational changes and CO2-induced channel gating in connexin26. Structure. https://doi.org/10.1016/j.str.2022.02.010
17. Qi C, Gutierrez SA, Lavriha P, Othman A, Lopez-Pigozzi D, Bayraktar E, Schuster D, Picotti P, Zamboni N, Bortolozzi M, Gervasio FL, Korkhov VM (2023) Structure of the connexin-43 gap junction channel in a putative closed state. Elife 12:RP87616. https://doi.org/10.7554/eLife.87616.3
18. Qi C, Lavriha P, Bayraktar E, Vaithia A, Schuster D, Pannella M, Sala V, Picotti P, Bortolozzi M, Korkhov VM (2023) Structures of wild-type and selected CMT1X mutant connexin 32 gap junction channels and hemichannels. bioRxiv 2023:2023–03. https://doi.org/10.1101/2023.03.08.531661
19. Buratto D, Donati V, Zonta F, Mammano F (2021) Harnessing the therapeutic potential of antibodies targeting connexin hemichannels. Biochim Biophys Acta Mol basis Dis 1867(4):166047. https://doi.org/10.1016/j.bbadis.2020.166047
20. Ribeiro JV, Bernardi RC, Rudack T, Stone JE, Phillips JC, Freddolino PL, Schulten K (2016) QwikMD—integrative molecular dynamics toolkit for novices and experts. Sci Rep 6:26536. https://doi.org/10.1038/srep26536
21. Jo S, Kim T, Iyer VG, Im W (2008) CHARMM-GUI: a web-based graphical user interface for CHARMM. J Comput Chem 29(11):1859–1865. https://doi.org/10.1002/jcc.20945
22. Case DA, Cheatham TE 3rd, Darden T, Gohlke H, Luo R, Merz KM Jr, Onufriev A, Simmerling C, Wang B, Woods RJ (2005) The Amber biomolecular simulation programs. J Comput Chem 26(16):1668–1688. https://doi.org/10.1002/jcc.20290
23. Pronk S, Pall S, Schulz R, Larsson P, Bjelkmar P, Apostolov R, Shirts MR, Smith JC, Kasson PM, van der Spoel D, Hess B, Lindahl E (2013) GROMACS 4.5: a high-throughput and highly parallel open source molecular simulation toolkit. Bioinformatics 29(7):845–854. https://doi.org/10.1093/bioinformatics/btt055
24. Brooks BR, Brooks CL 3rd, Mackerell AD Jr, Nilsson L, Petrella RJ, Roux B, Won Y, Archontis G, Bartels C, Boresch S, Caflisch A, Caves L, Cui Q, Dinner AR, Feig M, Fischer S, Gao J, Hodoscek M, Im W, Kuczera K, Lazaridis T, Ma J, Ovchinnikov V, Paci E, Pastor RW, Post CB, Pu JZ, Schaefer M, Tidor B, Venable RM, Woodcock HL, Wu X, Yang W, York DM, Karplus M (2009) CHARMM: the biomolecular simulation program. J Comput Chem 30(10):1545–1614. https://doi.org/10.1002/jcc.21287
25. Eastman P, Swails J, Chodera JD, McGibbon RT, Zhao Y, Beauchamp KA, Wang LP, Simmonett AC, Harrigan MP, Stern CD, Wiewiora RP, Brooks BR, Pande VS (2017) OpenMM 7: rapid development of high performance algorithms for molecular dynamics. PLoS Comput Biol 13(7):e1005659. https://doi.org/10.1371/journal.pcbi.1005659
26. Tian C, Kasavajhala K, Belfon KAA, Raguette L, Huang H, Migues AN, Bickel J, Wang Y, Pincay J, Wu Q, Simmerling C (2020) ff19SB: amino-acid-specific protein backbone parameters trained against quantum mechanics energy surfaces in solution. J Chem Theory Comput 16(1):528–552. https://doi.org/10.1021/acs.jctc.9b00591
27. Best RB, Zhu X, Shim J, Lopes PE, Mittal J, Feig M, Mackerell AD Jr (2012) Optimization of the additive CHARMM all-atom protein force field targeting improved sampling of the backbone phi, psi and side-chain chi(1) and chi(2) dihedral angles. J Chem Theory Comput 8(9):3257–3273. https://doi.org/10.1021/ct300400x
28. Barrera EE, Machado MR, Pantano S (2019) Fat SIRAH: coarse-grained phospholipids to explore membrane-protein dynamics. J Chem Theory Comput 15(10):5674–5688. https://doi.org/10.1021/acs.jctc.9b00435
29. Machado MR, Barrera EE, Klein F, Sonora M, Silva S, Pantano S (2019) The SIRAH 2.0 force field: Altius, Fortius, Citius. J Chem Theory

Comput 15(4):2719–2733. https://doi.org/10.1021/acs.jctc.9b00006

30. Pereira J, Simpkin AJ, Hartmann MD, Rigden DJ, Keegan RM, Lupas AN (2021) High-accuracy protein structure prediction in CASP14. Proteins 89(12):1687–1699. https://doi.org/10.1002/prot.26171
31. Jumper J, Evans R, Pritzel A, Green T, Figurnov M, Ronneberger O, Tunyasuvunakool K, Bates R, Zidek A, Potapenko A, Bridgland A, Meyer C, Kohl SAA, Ballard AJ, Cowie A, Romera-Paredes B, Nikolov S, Jain R, Adler J, Back T, Petersen S, Reiman D, Clancy E, Zielinski M, Steinegger M, Pacholska M, Berghammer T, Bodenstein S, Silver D, Vinyals O, Senior AW, Kavukcuoglu K, Kohli P, Hassabis D (2021) Highly accurate protein structure prediction with AlphaFold. Nature 596(7873):583–589. https://doi.org/10.1038/s41586-021-03819-2
32. Baek M, DiMaio F, Anishchenko I, Dauparas J, Ovchinnikov S, Lee GR, Wang J, Cong Q, Kinch LN, Schaeffer RD, Millan C, Park H, Adams C, Glassman CR, DeGiovanni A, Pereira JH, Rodrigues AV, van Dijk AA, Ebrecht AC, Opperman DJ, Sagmeister T, Buhlheller C, Pavkov-Keller T, Rathinaswamy MK, Dalwadi U, Yip CK, Burke JE, Garcia KC, Grishin NV, Adams PD, Read RJ, Baker D (2021) Accurate prediction of protein structures and interactions using a three-track neural network. Science 373(6557):871–876. https://doi.org/10.1126/science.abj8754
33. Johnstone SR, Billaud M, Lohman AW, Taddeo EP, Isakson BE (2012) Posttranslational modifications in connexins and pannexins. J Membr Biol 245(5–6):319–332. https://doi.org/10.1007/s00232-012-9453-3
34. Aasen T, Johnstone S, Vidal-Brime L, Lynn KS, Koval M (2018) Connexins: synthesis, post-translational modifications, and trafficking in health and disease. Int J Mol Sci 19(5). https://doi.org/10.3390/ijms19051296
35. Kwon T, Harris AL, Rossi A, Bargiello TA (2011) Molecular dynamics simulations of the Cx26 hemichannel: evaluation of structural models with Brownian dynamics. J Gen Physiol 138(5):475–493. https://doi.org/10.1085/jgp.201110679
36. Zonta F, Polles G, Sanasi MF, Bortolozzi M, Mammano F (2013) The 3.5 angstrom X-ray structure of the human connexin26 gap junction channel is unlikely that of a fully open channel. Cell Commun Signal 11(1):15. https://doi.org/10.1186/1478-811X-11-15
37. Zonta F, Mammano F, Torsello M, Fortunati N, Orian L, Polimeno A (2014) Role of gamma carboxylated Glu47 in connexin 26 hemichannel regulation by extracellular Ca(2)(+): insight from a local quantum chemistry study. Biochem Biophys Res Commun 445(1):10–15. https://doi.org/10.1016/j.bbrc.2014.01.063
38. Herrera FE, Pantano S (2009) Salt induced asymmetry in membrane simulations by partial restriction of ionic motion. J Chem Phys 130(19):195105. https://doi.org/10.1063/1.3132705
39. Zonta F, Buratto D, Crispino G, Carrer A, Bruno F, Yang G, Mammano F, Pantano S (2018) Cues to opening mechanisms from in Silico electric field excitation of Cx26 Hemichannel and in vitro mutagenesis studies in HeLa transfectans. Front Mol Neurosci 11:170. https://doi.org/10.3389/fnmol.2018.00170
40. Darre L, Machado MR, Brandner AF, Gonzalez HC, Ferreira S, Pantano S (2015) SIRAH: a structurally unbiased coarse-grained force field for proteins with aqueous solvation and long-range electrostatics. J Chem Theory Comput 11(2):723–739. https://doi.org/10.1021/ct5007746
41. Zonta F, Polles G, Zanotti G, Mammano F (2012) Permeation pathway of homomeric connexin 26 and connexin 30 channels investigated by molecular dynamics. J Biomol Struct Dyn 29(5):985–998. https://doi.org/10.1080/073911012010525027
42. Zonta F, Buratto D, Cassini C, Bortolozzi M, Mammano F (2014) Molecular dynamics simulations highlight structural and functional alterations in deafness-related M34T mutation of connexin 26. Front Physiol 5:85. https://doi.org/10.3389/fphys.2014.00085
43. Roux B, Karplus M (1991) Ion-transport in a gramicidin-like channel—dynamics and mobility. J Phys Chem 95:4856–4868
44. Torrie GMVJP (1977) Nonphysical sampling distributions in Monte Carlo free-energy estimation: umbrella sampling. J Comput Phys 23:187–199
45. Luo Y, Rossi AR, Harris AL (2016) Computational studies of molecular permeation through Connexin26 channels. Biophys J 110(3):584–599. https://doi.org/10.1016/j.bpj.2015.11.3528
46. Machado MR, Pantano S (2020) Split the charge difference in two! A rule of thumb for adding proper amounts of ions in MD simulations. J Chem Theory Comput 16(3):1367–1372. https://doi.org/10.1021/acs.jctc.9b00953

Chapter 5

Spatial and Temporal Localization of Connexins in Cells Using Confocal Microscopy

Sergiu A. Lucaciu, Stephanie E. Leighton, and Dale W. Laird

Abstract

The 21-member connexin family found in humans is the building block of both single-membrane spanning channels (hemichannels) and double-membrane spanning intercellular channels. These large-pore channels are dynamic and typically have a short life span of only a few hours. Imaging connexins from the time of synthesis in the endoplasmic reticulum through to their degradation can be challenging given their distinct assembly states and transient residences in many subcellular compartments. Here, we describe how connexins can be effectively imaged on a confocal microscope in living cells when tagged with fluorescent proteins and when immunolabeled with high affinity anti-connexin antibodies in fixed cells. Temporal and spatial localization of multiple connexins and disease-linked connexin mutants at the subcellular level extensively informs on the mechanisms governing connexin regulation in health and disease.

Key words Connexins, Gap junctions, Confocal microscopy, Imaging, Immunolabeling

1 Introduction

Visually tracking the life cycle of gap junction proteins (connexins) in cells has long been of interest owing in large part to their different states of assembly and their short half-life in most cell types. Historically, electron microscopic imaging together with immunogold labeling of connexins was in vogue up until the late twentieth century due to the exquisite subcellular details that could be resolved [1–3]. Today, this costly, static, and time-consuming means to assess connexin localization is rarely undertaken due in large part to the closure of many electron microscopy facilities as new and improved light microscopy technologies became available. Over four decades ago, light microscopy reached a new standard with the introduction of laser-driven confocal microscopy where the collection of fluorescent emissions from a single optical plane eliminated image distortion caused by out-of-focus light. With refinements in this technology that continued for over 40 years,

Fabio Mammano and Mauricio Retamal (eds.), *Connexin Hemichannels: Methods and Protocols*, Methods in Molecular Biology, vol. 2801, https://doi.org/10.1007/978-1-0716-3842-2_5,

access to confocal imaging is considered a "must have technology" in most cell biology research facilities today. While first-generation confocal microscopic imaging of immunolabeled connexins in fixed cells and tissues remains informative [4, 5], second-generation imaging of live cells engineered to express fluorescent protein-tagged connexins has revolutionized the understanding of connexin trafficking, assembly, and turnover in real-time [6–11]. Both of these imaging strategies provide important insights into connexin dynamics as connexin hexamers traffic to the cell surface to function as hemichannels or assembled into mature gap junctions where they facilitate intercellular communication [12]. Discovery that gene mutations in at least 11 of the 21 human connexin encoding genes leads to over 30 inherited diseases further increased the need to use imaging technologies to understand the fate of mutant proteins [13]. For example, imaging strategies revealed that many connexin-disease-linked mutants were trafficking defective and retained within subcellular compartments [14]. With the advent and enhancement of confocal microscopes equipped with multiple lasers for improved fluorochrome excitation and separation, it is now possible to optically separate three or more fluorescent signatures in 2-dimension (2D), 3D, or 4D with resolution well below 200 nm in x and y dimensions. These approaches continue to inform on how multiple coexpressed connexins and connexin mutants are sorted, assembled, and degraded in tissue-relevant and reference cell models establishing a necessary platform of knowledge in anticipation of future connexin-targeted therapeutics [15–19].

Here, we describe routinely used cell transfection, cell preparation, and image acquisition protocols for assessing temporal and spatial localization of connexins. While we feature the use of a Zeiss LSM 800 Airyscan confocal microscope and rat epidermal keratinocytes (REKs) [20, 21], protocols were adapted from our experience using earlier generation Zeiss LSM 410 and 510 META confocal imaging systems. REK cell culture, transfection, and preparation for imaging are relatively generic and have no direct bearing on what instrumentation will be used to perform the imaging. While the imaging specifications presented here are tailored for the Zeiss LSM 800 Airyscan confocal microscope, key concepts can be readily adapted to any current generation of Zeiss, Leica, or other confocal microscopes taking into account proprietary and instrument-specific specifications. Special attention will be given to localizing connexins during anterograde transport through intracellular compartments on route to the cell surface, at the cell surface, and retrograde transport to sites of degradation.

2 Materials

2.1 Cell Culture and Cell Transfection

All reagents are handled with aseptic technique principles in a sterile biosafety cabinet (*see* **Note 1**).

1. Dulbecco's Modified Eagle Medium (DMEM).
2. Stock solutions of fetal bovine serum (FBS); L-glutamine 200 mM (Gibco, cat# 25030-081), and penicillin (10,000 U/mL)/streptomycin (10,000 U/mL) are aliquoted in 50 mL or 15 mL conical centrifuge tubes and stored at −20 °C until use. Solutions are thawed on day of use.
3. Complete DMEM: In a sterile biosafety cabinet, supplement DMEM (without glutamine or pyruvate) with 10% FBS, 2 mM glutamine, and 50,000 U penicillin/streptomycin to a total volume of 500 mL. After mixing by inverting the closed bottle 4–6 times, the complete DMEM is warmed to 37 °C before using (*see* **Note 2**).
4. Stock solutions of trypsin/EDTA (0.25% trypsin, 2.2 mM EDTA tetrasodium) are stored in aliquots at −20 °C. The working aliquot is thawed and stored at 4 °C.
5. Sterile 25 cm^2 cell culture-treated flasks.
6. Sterile 60 mm tissue culture dishes.
7. 5¾″ borosilicate glass Pasteur pipette (autoclaved).
8. Borosilicate glass microscope coverslips (12 mm diameter).
9. OptiMEM reduced serum media (ThermoFisher, cat# 31985-070).
10. Lipofectamine™ 3000 transfection reagent (ThermoFisher, cat# L3000).

2.2 Cell Fixation and Labeling

1. Fixing solution: A mixed solution of 80% (v/v) methanol/20% (v/v) acetone is prepared in a fume hood and stored at −20 °C.
2. Sodium azide stock: Prepare a stock solution of 10% (w/v) sodium azide by dissolving 1 g of sodium azide in 10 mL of ddH_2O in a 15 mL conical centrifuge tube. Filter the solution through a 0.2 μm syringe filter and store at 4 °C.
3. Blocking solution: Add 1 mL phosphate-buffered saline (1× PBS, pH 7.4) to a 15 mL conical centrifuge tube. Add 0.2 g of bovine serum albumin (BSA) and top up to 10 mL with 1× PBS. Add 20 μL of the stock 10% sodium azide and vortex mixture (*see* **Note 3**). Store at 4 °C.
4. Hoechst 33342 stain: Tightly wrap a 15 mL conical centrifuge tube with aluminum foil to protect from light and add 10 mL ddH_2O and 10 μL Hoechst 33342 (16.2 mM). After vortexing, store at 4 °C.

5. Airvol mounting medium: Weigh 20 g of Airvol and transfer to a 100 mL graduated cylinder containing 20 mL of 1× PBS. Top to 58.8 mL with 1× PBS and stir the mixture for ~16 h at room temperature. Add 40 mL glycerol and 1.2 mL of 10% sodium azide to the Airvol/PBS solution and stir for an additional 16 h. Remove undissolved particles by centrifugation at 12,000 rpm for 25–30 min. Decant the supernatant and store in airtight vials or bottles.

2.3 Zeiss LSM 800 Airyscan Confocal Microscope

The Zeiss LSM 800 Airyscan confocal microscope was introduced in 2015 and has served to replace many LSM 510 META confocal microscopes. The LSM 800 inverted microscope is equipped with gallium arsenide phosphide photomultiplier tubes (GaAsP-PMTs) for high signal-to-noise detection of fluorescence, faster scan speeds (8–512 × 512 images/s) and an Airyscan detector that allows for 1.7× higher resolution image capture (up to 140 nm in x and y dimensions) than conventional confocal microscopes. The airy disk is composed of 32 single-detector elements that collect all signals from a sample that are reassigned to the correct position for optimal resolution and signal-to-noise ratio. The three detectors allow for three emission spectrums from live or fixed cells to be acquired simultaneously or sequentially and separated without overlapping signal bleed through and image contamination. The addition of a transmitted light detector increases the breadth of images that can be acquired. The system we use is equipped with three class 3B diode lasers that emit 405 (5 mW), 488 (10 mW), and 561 (10 mW) nm wavelengths that allows for the fluorescence emitted from most blue, green, and red fluorochromes and a wide variety of fluorescent proteins. The inverted microscope is fitted with a suite of objective lenses that includes a plan neofluar (10×, 0.3 NA (numerical aperture)), plan neofluar (25×, 0.8 NA), C-apochromat (40×, 1.2 NA, water), F fluar (40×, 0.8 NA, oil), and plan-apochromat (63×, 1.4 NA, oil) lenses. Images are captured on Zeiss Zen 2.3 software and archived on a connected server. In cases where images are obtained from live cells, the microscope stage can be fitted with a CTi controller 3700 digital system which delivers CO_2, adjustable heating via a Tempcontrol 37–2 controller and humidified air. Oil objectives can also be fitted with a temperature controlled mini heating collar (*see* **Note 4**).

3 Methods

3.1 Culturing Rat Epidermal Keratinocytes (REKs)

1. Warm complete DMEM in a 37 °C water bath for 15 min prior to use.
2. Culture REKs (*see* **Note 5**) in sterile 25 cm^2 cell culture-treated flasks or 60 mm tissue culture dishes that are housed in an

incubator maintained at 37 °C and 5% CO_2. Passage REKs once they reach an estimated confluence of ~90% (*see* **Note 6**).

3. In preparation for subculturing cells, remove the 25 cm^2 flask of REKs from the incubator and check cell viability (i.e., no sign of floating cells, blebbing) and confluence under a light microscope. Transfer the flask of REKs into the biosafety cabinet.
4. Attach an autoclaved 5¾″ borosilicate glass Pasteur pipette to a vacuum line and aspirate the growth medium from the flask.
5. Gently rinse the cell monolayer three times with 1–2 mL of sterile 1× PBS (pH 7.4). Use the Pasteur pipette to aspirate 1× PBS between washes.
6. Warm an aliquot of trypsin/EDTA to 37 °C. Aspirate the final volume of 1× PBS and gently add 1 mL trypsin/EDTA to cover the cell surface. Return the flask to the incubator for up to 10 min.
7. Once 7 min have elapsed, check for cell detachment under a light microscope. If the cells are not detached, return to the incubator, and check every minute until the cells have visibly detached from the bottom of the flask.
8. In the biosafety cabinet add 1 mL of complete DMEM to inhibit the trypsin. Gently mix the suspension by pipetting up and down to break apart any cell aggregates.
9. Transfer the 2 mL cell suspension to a sterile 15 mL conical centrifuge tube.
10. Pellet the cells by centrifugation at 1000–1500 rpm (~180–300 g) for 3–5 min using a swinging bucket tabletop centrifuge.
11. Return the conical tube to the sterile biosafety cabinet and aspirate the supernatant. Take extra care to avoid disrupting the cell pellet.
12. Using a P1000 pipette, gently resuspend and disrupt the cell pellet in 1 mL complete DMEM.
13. Top up the cell suspension to the desired final volume (usually 5 mL) with complete DMEM (*see* **Notes** 7 and **8**).
14. When cells being cultured are destined to be used for confocal imaging, REKs are cultured in 60 mm dishes that contain glass coverslips. Borosilicate glass microscope coverslips (12 mm diameter) are stored in 95% ethanol. Place as many coverslips as needed into each dish and allow the ethanol to evaporate prior to adding 3 mL of culture medium (*see* **Note 9**).
15. Transfer 1 mL of the cell suspension to prepare 60 mm dishes (*see* **Note 10**). Gently rock the dish side-to-side to distribute cells throughout the dish and place within the incubator for

24 h. As an alternative, trypsinized cells can be counted and plated at a seeded concentration of ~0.3–2.0 × 10^5 cells/mL.

3.2 Transfecting REKs

1. Warm complete DMEM and OptiMEM reduced serum media in a 37 °C water bath.
2. Check cells under a light microscope to ensure cell confluence is between 50% and 70%.
3. Aspirate the growth medium from the flask.
4. Gently rinse the cell monolayer 3× with 1–2 mL of sterile 1× PBS (pH 7.4).
5. Replenish cells with 4 mL complete DMEM and return to incubator.
6. Label two 1.5 mL microcentrifuge tubes (tube A, tube B) for each transfection reaction.
7. Thaw aliquots of plasmid DNA at room temperature, and place in biosafety cabinet along with Lipofectamine 3000 and P3000 reagent tubes.
8. For each transfection reaction, prepare tube A to contain a mixture of 125 μL warm OptiMEM and 3.75 μL Lipofectamine™ 3000 Reagent. Prepare tube B to contain a mixture of 125 μL of warm OptiMEM and a volume equivalent to 1 μg plasmid DNA (*see* **Note 11**) and 2 μL P3000™ Enhancer Reagent for every 1 μg of plasmid DNA. Mix gently by pipetting. Each transfection reaction is used for one dish of cells.
9. Add contents of tube B to tube A and mix gently by pipetting. Incubate for 5 min at room temperature.
10. Retrieve 60 mm dishes containing the REKs to be transfected from the incubator and place in the biosafety cabinet.
11. Add the transfection solution dropwise onto the cells throughout the 60 mm cell culture dish.
12. Gently rock the culture dish side-to-side to distribute the DNA–lipofectamine complexes throughout the cell culture dish. Return the cells to the incubator for 24–48 h.

3.3 Fixing and Immunolabeling Cells

1. Check and record viability of REKs after transfection (i.e., no sign of floating cells, blebbing) and estimate confluence under a light microscope.
2. In the biosafety cabinet, aspirate the growth medium from the flask.
3. Gently rinse the cell monolayer three times with 1–2 mL of sterile 1× PBS (pH 7.4).
4. Submerge cells in ice-cold fixing solution at 4 °C for 15 min.

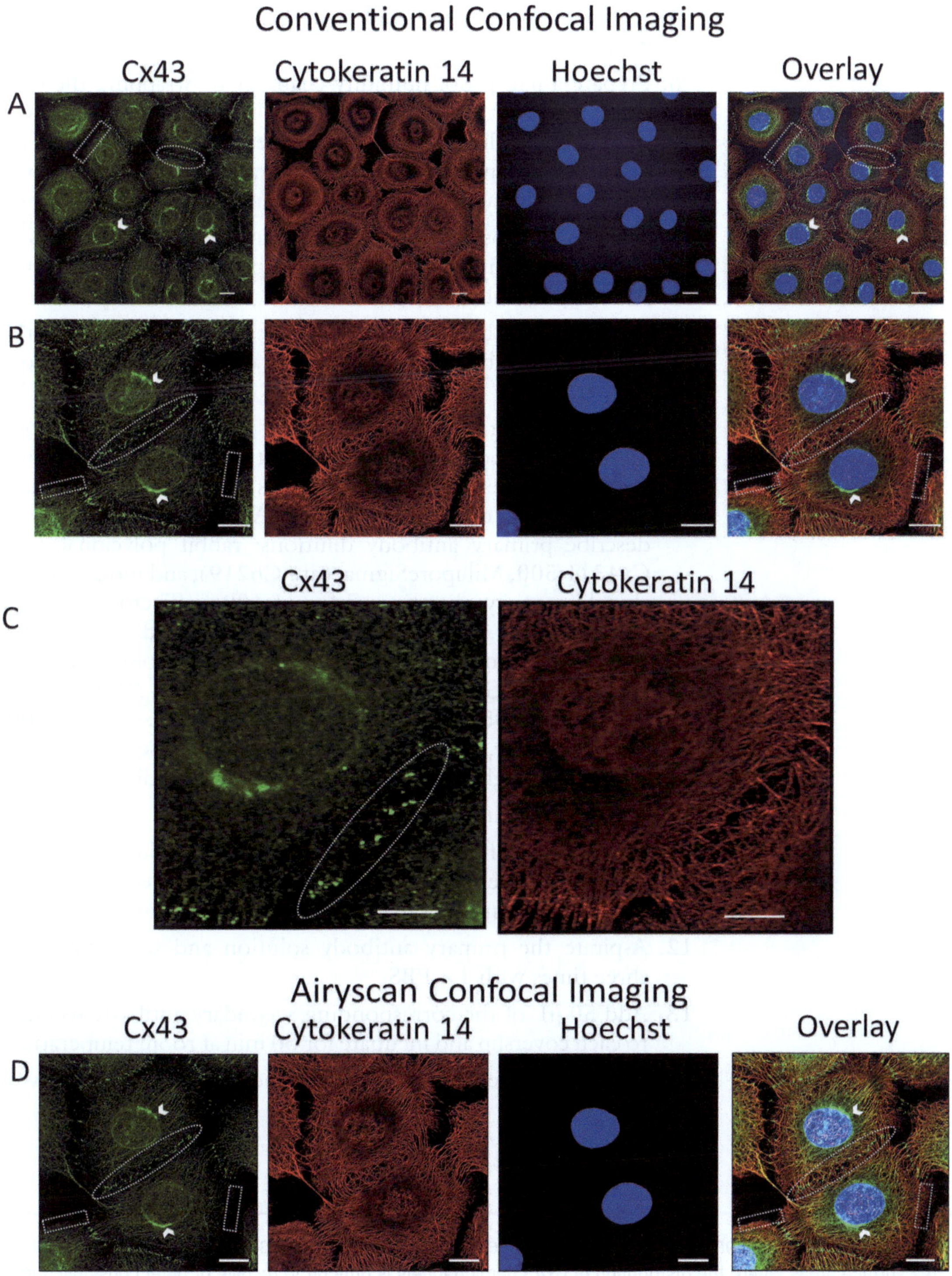

Fig. 1 Confocal imaging of immunolabeled Cx43 and cytokeratin 14 found in REKs. Low (**a**), high (**b**), and higher (**c**) magnification images depict the localization of Cx43 (green) with respect to the distribution of cytokeratin 14 intermediate filaments (red). Airyscan confocal image (**d**) of the identical ROI shown in (**b**) but acquired at a slightly different optical plane. The digital ovals demarcate the typical display of Cx43 gap junctions at the surface of two contacting cells. Note the variability of gap junction size reflecting that gap

5. Aspirate fixing solution and wash three times with 1× PBS (*see* **Note 12**).
6. Prepare a lightproof humidity chamber by wrapping the outsides of a 10 cm dish and its lid with aluminum foil. Cut a piece of Whatman filter paper to fit inside the dish and soak with ddH_2O. Cut one or two pieces of parafilm and place on top of the wet filter paper. Cover with the lid (*see* **Note 13**).
7. While coverslips are still submerged in 1× PBS, use forceps to transfer coverslips with the cell side up onto the parafilm in the humidity chamber.
8. Add 50 μL of blocking solution to each coverslip. Cover the humidity chamber with the lid and incubate for 30 min at room temperature.
9. While blocking, prepare the primary and secondary antibody dilutions in blocking solution as per the antibody information sheet or from experiential learning as to the most appropriate concentration or dilution to use. As an example, here we describe primary antibody dilutions: rabbit polyclonal anti-Cx43 (1:500, MilliporeSigma, cat# C6219), and mouse monoclonal anti-cytokeratin 14 (1:100, Thermo Fisher MA5-11599) (*see* Figs. 1, 2). Prepare corresponding secondary antibody dilutions as follows: Alexa Fluor 488-conjugated goat anti-rabbit IgG (1:500, Thermo Fisher, cat# A11008) and Alexa Fluor 555-conjugated goat anti-mouse IgG (1:500, Thermo Fisher, cat# A32727). Protect diluted secondary antibodies from light and store dilutions at 4 °C until ready to use.
10. Aspirate blocking solution.
11. Add 50 μL of the first primary antibody solution to each coverslip. Cover the humidity chamber with the lid and incubate for 60 min at room temperature.
12. Aspirate the primary antibody solution and wash coverslips three times with 1× PBS.
13. Add 50 μL of the corresponding secondary antibody solution to each coverslip and incubate for 60 min at room temperature.
14. Aspirate the secondary antibody and wash three times with 1× PBS.

Fig. 1 (continued) junctions can be composed of a few to hundreds of channels. The rectangular boxes depict free surface areas where the distribution of Cx43 hemichannels is difficult to resolve or detect possibly due to their low abundance. Chevrons denote paranuclear sites of Cx43 reminiscent of the Golgi apparatus. The 16-bit Airyscan images acquired with a 63× oil objective are large, require ~4–8× longer to acquire, and have a 1.7× better resolution compared to conventional confocal images. Hoechst staining of nuclei is denoted in blue. Bars = 10 μm

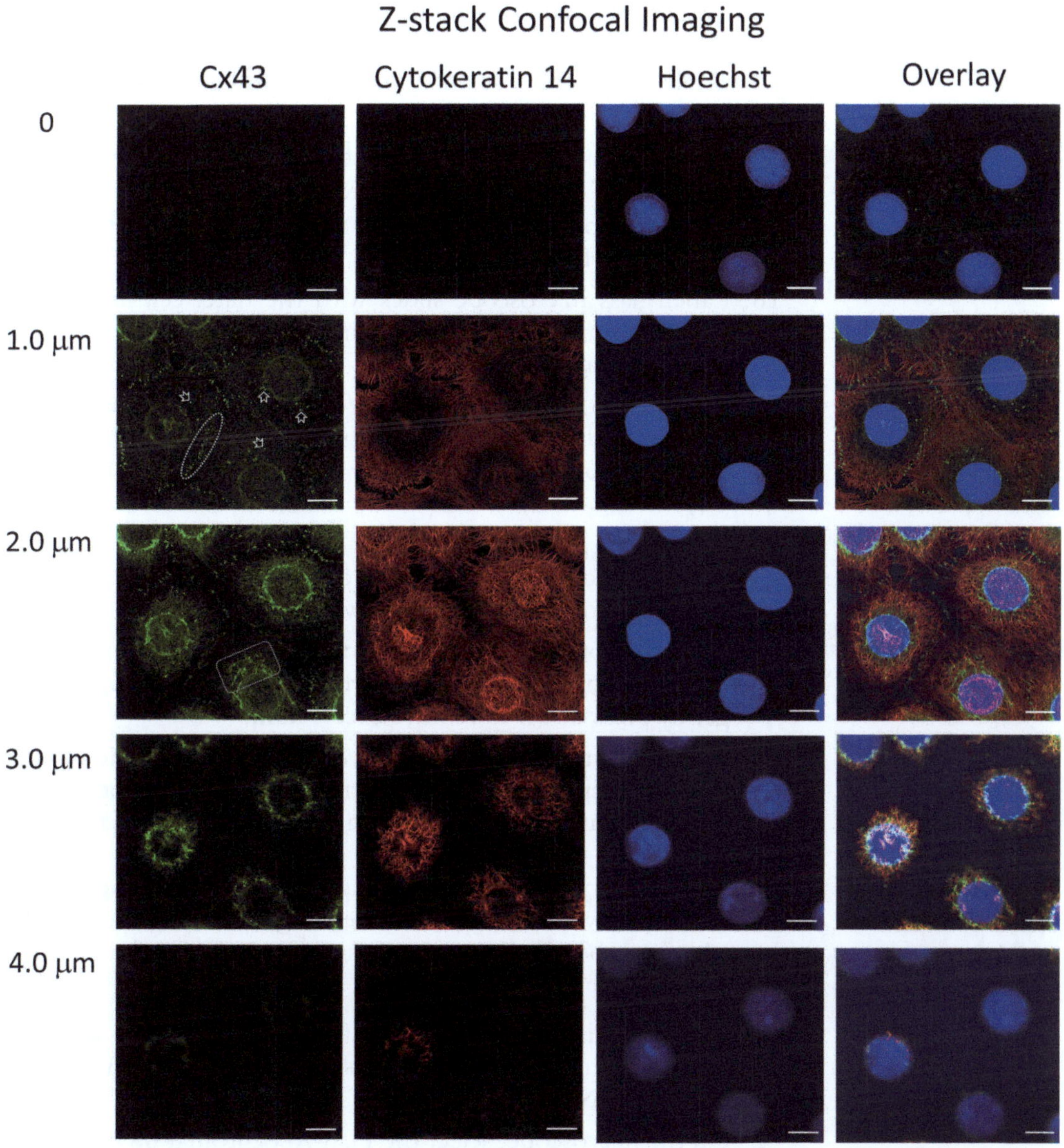

Fig. 2 Optical sectioning of REKs immunolabeled for Cx43 and cytokeratin 14. A total of five optical slices in the *z* dimension were acquired using a 63× oil objective with a set interval of 1.00 μm in order to capture the distribution of Cx43 (green), cytokeratin 14 (red), and Hoechst (blue). Each optical plane revealed a distinct distribution pattern for Cx43 and cytokeratin reflecting their spatial location within REKs. Arrows denote Cx43-containing vesicular structures reflecting Cx43 transiently captured within anterograde or retrograde subcellular compartments. Dashed box highlights Cx43 residing within the endoplasmic reticulum or tubular structures that may play a role in antegrade Cx43 delivery to the cell surface. Confocal imaging series was acquired over a period of 8.5 min. Bars = 10 μm

15. To double label, repeat steps 11–14 with the second set of primary and secondary antibody solutions (i.e., mouse anti-cytokeratin 14, and Alexa Fluor 555-conjugated anti-mouse IgG).
16. Add 50 μL of Hoechst 33342 (1:1000) to each coverslip and incubate for 1 min at room temperature to stain the nuclei.
17. Aspirate Hoechst 33342 solution off the coverslip and add 100 μL ddH_2O.
18. Label glass microscope slides with the date, cell line, transfected plasmid (if applicable), immunolabeled proteins, and secondary antibodies.
19. Add 1 drop of Airvol mounting medium on the glass microscope slide for each coverslip. Ensure the drops of mounting media are spaced out sufficiently to fit the coverslips (*see* **Note 14**).
20. Using forceps, carefully lift the coverslip from the humidity chamber and gently dab the side of the coverslip onto a piece of filter paper or Kimwipe to remove excess ddH_2O.
21. Place the coverslip cell-side-down onto the drop of mounting medium. Invert slide with coverslip onto paper towel and gently press down to remove excess fluid.
22. Place glass microscope slides coverslip-side-up in a slide folder at 4 °C. Prepared slides maintained in the dark can be imaged many days after preparation with minimal photobleaching.

3.4 Single Slice Optical Confocal Imaging

The vast majority of confocal imaging involves single image acquisition of one, two, or three fluorescent signals. Here, we describe how these can be obtained on a Zeiss LSM 800 Airyscan confocal microscope.

1. Image acquisition is achieved using the ZEISS ZEN 2.3 software platform installed on a Zeiss LSM 800 Airyscan imaging system (*see* **Note 15**).
2. On the microscope, select the Plan-Apochromat 63×/1.4 NA oil immersion objective lens (*see* **Note 16**).
3. Tilt the microscope head back to access the objective lenses and apply one drop of immersion oil onto the objective.
4. Place the glass microscope slide gently onto the mechanical stage with the coverslip directed above the objective lens. Use the coarse and fine adjustment knob to raise the lens objective into contact with the immersion oil. Return the microscope head to the upright position.
5. Using the ZEISS ZEN 2.3 software, select the locate tab and a wavelength for focusing (*see* **Note 17**). Once selected, use the speedDIAL remote to turn on the LED illumination.

6. While looking through the ocular lenses, focus on the specimen and use the joystick controller to orient the field of view and locate a region of interest (ROI).
7. Select the acquisition tab to switch to the laser scanning settings and select an imaging protocol from the dropdown menu (*see* **Note 18**).
8. As an example of parameters commonly used, set image capture to 16 bit, a scan speed of 6, image averaging of 4, and a zoom of 1.5 (*see* **Note 19**).
9. Select the imaging channels if not already preassigned in the chosen imaging protocol (*see* **Note 20**).
10. Under the acquisition tab, select the live button. Adjust the fine focus to emphasize connexins found in subcellular organelles, vesicles, cell surface distributions, and gap junction plaques. This can also be accomplished by selecting the control key on the keyboard and adjusting with the mouse wheel.
11. Return to the channel window and adjust the laser strength to minimize pixel saturation (*see* **Note 21**) and select optimal frame size and master gain for image acquisition. Laser intensity should be kept to a minimum (~2%) to avoid fluorochrome photobleaching. In cases where images are to be quantified and compared within an experimental set, identical settings need to be used to acquire all images.
12. To acquire and save an image (CZI format), select the snap button under the acquisition tab (*see* **Note 22**).
13. Export single or multiple images as a BMP, TIFF, PNG, WDP, or SUR file (*see* **Note 23**). Export all channel collected images to the desired destination folder.

3.5 Optical Sectioning in Z-Dimension

Z-dimensional optical imaging and image stacking are commonly used to portray a representation of connexin distribution in *x*, *y*, and *z* dimensions in cases where connexins are found at multiple optical planes that can exist even in monolayer cultures of cells. For example, depending on whether REK monolayers are fixed in methanol/acetone or paraformaldehyde, they will exhibit a z-thickness of 5–15 μm allowing for many optical images to be collected and compiled.

1. Follow **steps 1–7** for single slice optical imaging (*see* **Note 24**).
2. Under the acquisition tab, check the z-stacking box, advance to the z-stack window and select the live button. Then use the fine focus to capture the upper boundary of the cell as defined by evidence of diminishing fluorescence (*see* **Note 25**). Select the set first button.

3. Adjust the focus to identify the opposite boundary (base) of the cell where minimal fluorescence will be detected. Select the set last button.
4. Adjust the number of optical slices to be captured (*see* **Note 26**).
5. Under the acquisition tab, unselect the live button and then select start experiment.
6. Acquired images are saved and exported as described above.
7. Before export, z-section acquired images are stacked/flattened for presentation using the extended depth of focus or orthogonal projection processing features, both found within the ZEISS ZEN 2.3 software, using Maximum Projection option from the method dropdown list.

3.6 Airyscan

Airyscan is used when the desire is to achieve the best resolution, signal-to-noise ratio, and image clarity possible to interrogate the finer details of connexin subcellular localization. This mode of imaging is considerably time consuming but improves *x/y* image resolution to ~140 nm when using a 63×, 1.4 NA, oil objective lens. It should be noted that image colors will appear dimmer as the Airyscan detector resolves more pixels that lack fluorescent signals as signal only originates from tiny discrete points where the fluorophores are located.

1. Follow **steps 1–7** for single slice optical imaging.
2. Using the ZEISS ZEN 2.3 software, select the acquisition tab to switch to the laser scanning settings. Then select an imaging protocol from the dropdown menu (*see* **Note 27**).
3. Under the acquisition tab, advance to the acquisition mode parameters and adjust to 16 bit images, average of 2, zoom of 1.3×, and a scan speed of 4 (*see* **Note 28**).
4. Select optimal frame size.
5. Follow **steps 9–11** for single slice optical imaging.
6. Select Airyscan processing and then select the corresponding Airyscan file as input and apply. A new document with the processed image will appear.
7. Save processed images as CZI files (*see* **Note 29**).
8. Processed images can be exported as described earlier.

3.7 Live Cells Imaging of Fluorescent Protein-Tagged Connexins

REKs and other mammalian cells can be engineered to express fluorescent protein-tagged connexins where the tag is added to the C-terminal of the connexins (*see* **Note 30**) or to the N-terminal of the connexin (*see* **Note 31**).

1. For high-quality and high-magnification imaging using the 63× objective lens, cells need to be cultured directly in 35 mm glass-bottom dishes or grown on glass-coverslips and placed carefully cell-side-down on 35 mm glass-bottomed dishes. Cells engineered to express fluorescent protein-tagged connexins should be imaged 24–72 h after transfection and prior to the cells overgrowing.
2. Place cells in fresh OptiMEM held at 37 °C using the CO_2-fed heated stage positioned on top of the confocal motorized stage that has a heated 63× oil objective. Bathe cells in 5% CO_2 and humidified air to maintain pH and temperature control (*see* **Note 32**).
3. ROI should be identified by visually using the oculars and *x*, *y*, and *z* manipulations to scan the field of cells.
4. Localize the ROI to the center of view and quickly switch the instrument to imaging mode to avoid photobleaching.
5. Optimal image quality for a time-lapse series is determined by adjusting the pinhole diameter, laser intensity, scan speed, contrast and brightness settings to the desired levels as described for single slice optical imaging (*see* **Note 33**).
6. Image acquisition is set to a desired speed and optimal frame size with slower speeds providing enhanced signal-to-noise ratios at the expense of possible photobleaching and longer acquisition time.
7. In cases where the connexin is tagged with a green fluorescent protein, cells are excited using the 488 nm laser with the resulting emitted fluorescence acquired on an assigned GaAsP PMT set to capture emissions within 500–550 nm. For connexins tagged with a red fluorescent protein, cells are excited with a 561 nm laser with emissions collected on a GaAsP PMT assigned to capture emissions between 600 and 655 nm.
8. Using the Zeiss software, set the time-lapse image acquisition parameters which include the number of images, time intervals between image acquisition, and total time of imaging (*see* **Note 34**).
9. Images are stored as previously described and animated as movies using either built-in Zeiss or other image processing software.

4 Notes

1. Prior to using the biosafety cabinet, turn on the fans for 15 min and spray down all accessible surfaces within the cabinet and items placed in the cabinet with 70% ethanol (or 70% isopropanol).

2. If L-glutamine and/or penicillin–streptomycin aliquots precipitate during thawing, dissolve by vortexing.
3. May need to vortex for several minutes to completely dissolve the BSA powder.
4. The methods described here to image and localize connexins in fixed and live cells were originally developed and refined from our experiential knowledge of using our current instrument and previous generations of Zeiss microscopes (LSM 410 and LSM 510 META systems) [6–8, 22–26]. It should be possible to adapt the methods described here to all currently used confocal microscope systems.
5. Procedures presented can be readily adapted to culture a wide variety of mammalian cell types.
6. Do not allow REKs to exceed 100% confluence as overgrowing will cause the cells to stratify and begin to spontaneously differentiate.
7. REKs will double in ~24 h. A 1:5 dilution of cells taken from a 90% confluent 25 cm^2 flask of REKs will result in a cell seeding confluence in a 60 mm dish of ~20%. Seeded REKs will grow to a confluence of ~80% in ~48 h.
8. A cell confluence of 50–70% is optimal for transfection reactions using the Lipofectamine 3000 kit. This can be achieved in a 60 mm dish after a 24-h incubation from a 1:3 or 1:4 dilution from a ~ 90% confluent 25 cm^2 flask of cells. Additional or larger dishes may be required depending on the number of dishes of cells being prepared.
9. It is good practice to include extra coverslips in each dish in the event a coverslip becomes damaged, or a late decision is made to expand the number of antibodies used for immunolabeling.
10. To ensure consistent seeding across multiple dishes, mix the cell suspension thoroughly and pipette cells from the middle of the cell suspension.
11. Many of the connexin, mutant connexin, and fluorescent protein-tagged connexin encoding pcDNA plasmids are driven from the cytomegalovirus enhancer-promoter. Typically, we find that 0.5–1 μg of DNA is ideal for cells grown in 35 mm or 60 mm culture dishes.
12. Fixed cells can be stored at 4 °C submerged in 1× PBS for up to 2 weeks, although optimal results are achieved if cells are immunolabeled shortly after fixation. As a cell fixation alternative, cells can be fixed in 4% paraformaldehyde (w/v)/PBS for 20 min followed by treating with 0.2% Triton X-100/PBS to ensure the membranes are permeable for antibody labeling protocols.

13. A small amount of 1× PBS should remain covering the cells which will assist in distributing the blocking solution over the entire coverslip and prevent droplets from forming.
14. You can comfortably fit three 12 mm glass coverslips on a standard glass microscope slide.
15. Zeiss provides periodic updates to the Zen software.
16. While many objective lenses can be chosen, the use of a high-magnification objective, such as the Plan-Apochromat 63×/1.4 NA oil immersion lens, enables the capture of high-resolution images that clearly resolve connexins in multiple subcellular locations.
17. It is best to use the brightest fluorescent signal in the sample for focusing.
18. Most imaging systems can be programmed to have several preset imaging protocols that activate the desired lasers and set the image acquisition parameters for each detector.
19. Conditions for image capture are customized pending the strength of the fluorescent signals and the desired signal-to-noise ratio. For example, increasing averaging to 4 greatly improves the signal-to-noise ratio.
20. Since we often double-label REKs and stain nuclei with Hoechst, three channels for consecutive image acquisition are selected (405 nm for detecting Hoechst 33342, 488 nm for detecting GFP or Alexa Fluor 488, and 561 nm for detecting RFP or Alexa Fluor 555).
21. To visualize pixel saturation, select the range indicator box. Saturated pixels will appear red in color. Aim for minimal pixel saturation by ensuring only the regions with the highest signal appear red/saturated. If images are to be quantified for signal intensity, it is important to ensure images are not saturated. The laser strength and master gain can be manually adjusted to minimize pixel saturation.
22. We recommend capturing approximately 10–25 images representative images per sample to allow for future quantification if desired.
23. We typically export images as TIFF or PNG files.
24. To minimize sample movement when using oil objective lenses, ensure the slide or glass-bottomed dish is placed on the microscope stage 5–10 min before imaging.
25. In the z-stack window within the Zen software, the location relative to the objective is visually depicted.
26. For monolayer cultures of REKs, we typically collect four to five images where each optical section captures ~1.2 μm of optical fluorescence minimizing optical plane overlap.

27. The Zeiss LSM 800 microscope has an Airyscan protocol to select.
28. Imaging conditions can be adjusted by the user to meet the desired image quality and speed of acquisition.
29. Due to the large size of the files, we do not recommend saving the unprocessed Airyscan file but rather suggest saving the processed CZI files.
30. There are now many variants of fluorescent proteins that acquire a molecular structure that establishes a spectrum of chromophores at the core. Molecular engineering fluorescent protein variants have further yielded a wide color palette of monomeric fluorescent proteins that have enhanced brightness and resistance to unintentional photobleaching. Surprisingly, these protein tags are largely benign (or have known limitations) when attached to the C-terminal of most connexins allowing the life cycle of connexins to be followed in living cells [6–8, 22–26].
31. Molecular tagging of fluorescent proteins to the N-terminal of connexins inhibits channel function of many tested connexins but still allows the formation of gap junction plaques at sites of cell–cell apposition [7]. Fortuitously, this tagging approach also changes the kinetics of connexin assembly into gap junctions and allows for connexons/hemichannels to be easily detected at the plasma membrane prior to docking with channels from apposed cells [8].
32. In cases where the confocal imaging system to be used is not equipped with a CO_2-fed heated stage, it is possible to perform short duration live time-lapse experiments. Here, cells are switched to 37 °C culture medium containing 10 mM HEPES to maintain the pH at 7.2. Live cell imaging should be limited to <10 min taking into account that gradual cooling of the medium could lead to temperature-induced changes to the life cycle of connexins.
33. It is particularly important to keep the laser intensity to a minimum to avoid photobleaching effects that may be seen when a time-lapse series is presented in tile or movie format.
34. In cases where connexin vesicles are being tracked, images should be acquired every 1 s due to their rapid rate of movement. Assessment of gap junction plaque movement or fate can be tracked by acquiring images every 10–60 s. In all cases, the frequency of image acquisition needs to take into account the potential of photobleaching the specimen.

Acknowledgments

The authors thank Kevin Barr for his valuable input. This work was supported by the Canadian Institutes of Health Research (grants 148630 and 148584).

References

1. Johnson R, Hammer M, Sheridan J, Revel JP (1974) Gap junction formation between reaggregated Novikoff hepatoma cells. Proc Natl Acad Sci USA 71:4536–4540
2. Caspar DL, Goodenough DA, Makowski L, Phillips WC (1977) Gap junction structures. I. Correlated electron microscopy and x-ray diffraction. J Cell Biol 74:605–628
3. Dermietzel R, Yancey B, Janssen-Timmen U, Traub O, Willecke K, Revel JP (1987) Simultaneous light and electron microscopic observation of immunolabeled liver 27 KD gap junction protein on ultra-thin cryosections. J Histochem Cytochem 35:387–392
4. Gourdie RG, Green CR, Severs NJ (1991) Gap junction distribution in adult mammalian myocardium revealed by an anti-peptide antibody and laser scanning confocal microscopy. J Cell Sci 99:41–55
5. Laird DW, Yancey SB, Bugga L, Revel JP (1992) Connexin expression and gap junction communication compartments in the developing mouse limb. Dev Dyn 195:153–161
6. Jordan K, Solan JL, Dominguez M, Sia M, Hand A, Lampe PD, Laird DW (1999) Trafficking, assembly, and function of a connexin43-green fluorescent protein chimera in live mammalian cells. Mol Biol Cell 10: 2033–2050
7. Laird DW, Jordan K, Thomas T, Qin H, Fistouris P, Shao Q (2001) Comparative analysis and application of fluorescent protein-tagged connexins. Microsc Res Tech 52:263–272
8. Simek J, Churko J, Shao Q, Laird DW (2009) Cx43 has distinct mobility within plasma-membrane domains, indicative of progressive formation of gap-junction plaques. J Cell Sci 122:554–562
9. Falk MM, Lauf U (2001) High resolution, fluorescence deconvolution microscopy and tagging with the autofluorescent tracers CFP, GFP, and YFP to study the structural composition of gap junctions in living cells. Microsc Res Tech 52:251–262
10. Fong JT, Kells RM, Gumpert AM, Marzillier JY, Davidson MW, Falk MM (2012) Internalized gap junctions are degraded by autophagy. Autophagy 8:794–811
11. Fort AG, Murray JW, Dandachi N, Davidson MW, Dermietzel R, Wolkoff AW, Spray DC (2011) In vitro motility of liver connexin vesicles along microtubules utilizes kinesin motors. J Biol Chem 286:22875–22885
12. Laird DW (2006) Life cycle of connexins in health and disease. Biochem J 394:527–543
13. Laird DW, Lampe PD (2022) Cellular mechanisms of connexin-based inherited diseases. Trends Cell Biol 32:58–69
14. Yum SW, Kleopa KA, Shumas S, Scherer SS (2002) Diverse trafficking abnormalities of connexin32 mutants causing CMTX. Neurobiol Dis 11:43–52
15. Xu L, Carrer A, Zonta F, Qu Z, Ma P, Li S, Ceriani F, Buratto D, Crispino G, Zorzi V et al (2017) Design and characterization of a human monoclonal antibody that modulates mutant Connexin 26 Hemichannels implicated in deafness and skin disorders. Front Mol Neurosci 10:298
16. Srinivas M, Verselis VK, White TW (2018) Human diseases associated with connexin mutations. Biochim Biophys Acta 1860:192–201
17. Retamal MA, Fernandez-Olivares A, Stehberg J (2021) Over-activated hemichannels: a possible therapeutic target for human diseases. Biochim Biophys Acta Mol basis Dis 1867:166232
18. Laird DW, Lampe PD (2018) Therapeutic strategies targeting connexins. Nat Rev Drug Discov 17:905–921
19. Kuang Y, Zorzi V, Buratto D, Ziraldo G, Mazzarda F, Peres C, Nardin C, Salvatore AM, Chiani F, Scavizzi F et al (2020) A potent antagonist antibody targeting connexin hemichannels alleviates Clouston syndrome symptoms in mutant mice. EBioMedicine 57: 102825
20. Baden HP, Kubilus J (1983) The growth and differentiation of cultured newborn rat keratinocytes. J Invest Dermatol 80:124–130
21. Maher AC, Thomas T, Riley JL, Veitch G, Shao Q, Laird DW (2005) Rat epidermal keratinocytes as an organotypic model for

examining the role of Cx43 and Cx26 in skin differentiation. Cell Commun Adhes 12:219–230

22. Jordan K, Chodock R, Hand AR, Laird DW (2001) The origin of annular junctions: a mechanism of gap junction internalization. J Cell Sci 114:763–773
23. Laird DW, Jordan K, Shao Q (2001) Expression and imaging of connexin-GFP chimeras in live mammalian cells. Methods Mol Biol 154: 135–142
24. Thomas T, Jordan K, Simek J, Shao Q, Jedeszko C, Walton P, Laird DW (2005) Mechanisms of Cx43 and Cx26 transport to the plasma membrane and gap junction regeneration. J Cell Sci 118:4451–4462
25. Au A, Shao Q, White KK, Lucaciu SA, Esseltine JL, Barr K, Laird DW (2020) Comparative analysis of Cx31 and Cx43 in differentiation-competent rodent keratinocytes. Biomol Ther 10(10):1443
26. Lucaciu SA, Shao Q, Figliuzzi R, Barr K, Bai D, Laird DW (2022) Interrogation of carboxy-terminus localized GJA1 variants associated with erythrokeratodermia variabilis et progressiva. Int J Mol Sci 23(1):486

Chapter 6

Assessment of Connexin43 Hemichannel Functionality Based on Cytosolic Uptake of Yo-Pro1

Raf Van Campenhout, Julen Sanz-Serrano, Axelle Cooreman, Yannick De Vlaeminck, Karine Breckpot, Prashant Kadam, Andrés Tabernilla, and Mathieu Vinken

Abstract

Connexin proteins are the building blocks of gap junctions and connexin hemichannels. Both provide a pathway for cellular communication. Gap junctions support intercellular communication mechanisms and regulate homeostasis. In contrast, open connexin hemichannels connect the intracellular compartment and the extracellular environment, and their activation fuels inflammation and cell death. The development of clinically applicable connexin hemichannel blockers for therapeutic purposes is therefore gaining momentum. This chapter describes a well-established protocol optimized for assessing connexin hemichannel activity by using the reporter dye Yo-Pro1.

Key words Cellular communication, Connexin proteins, Connexin43 hemichannel functionality, Lentiviral transduction of connexin43, Cytosolic uptake of Yo-Pro1

1 Introduction

Connexin (Cx) proteins are expressed by many types of cells across multiple tissues. As much as 21 different Cx species have been characterized in the human body, all named based on their molecular weight [1, 2]. In this regard, cDNA sequencing of the Cx43 gene product indicates a molecular mass of 43 kDa for this protein [3]. Cx proteins are the building blocks of cellular communication channels. Upon translation, six Cx proteins assemble to form Cx hemichannels. These Cx hemichannels build gap junctions through the docking of Cx hemichannels of adjacent cells. Gap junctions establish a pathway for intercellular communication by allowing the diffusion of small (1–1.5 kDa) hydrophilic molecules (glucose, glutamate, glutathione, adenosine triphosphate, cyclic adenosine monophosphate, inositol triphosphate), and ions (calcium, sodium, and potassium) across the cell plasma membrane [4–7]. Besides

Fabio Mammano and Mauricio Retamal (eds.), *Connexin Hemichannels: Methods and Protocols*, Methods in Molecular Biology, vol. 2801, https://doi.org/10.1007/978-1-0716-3842-2_6,

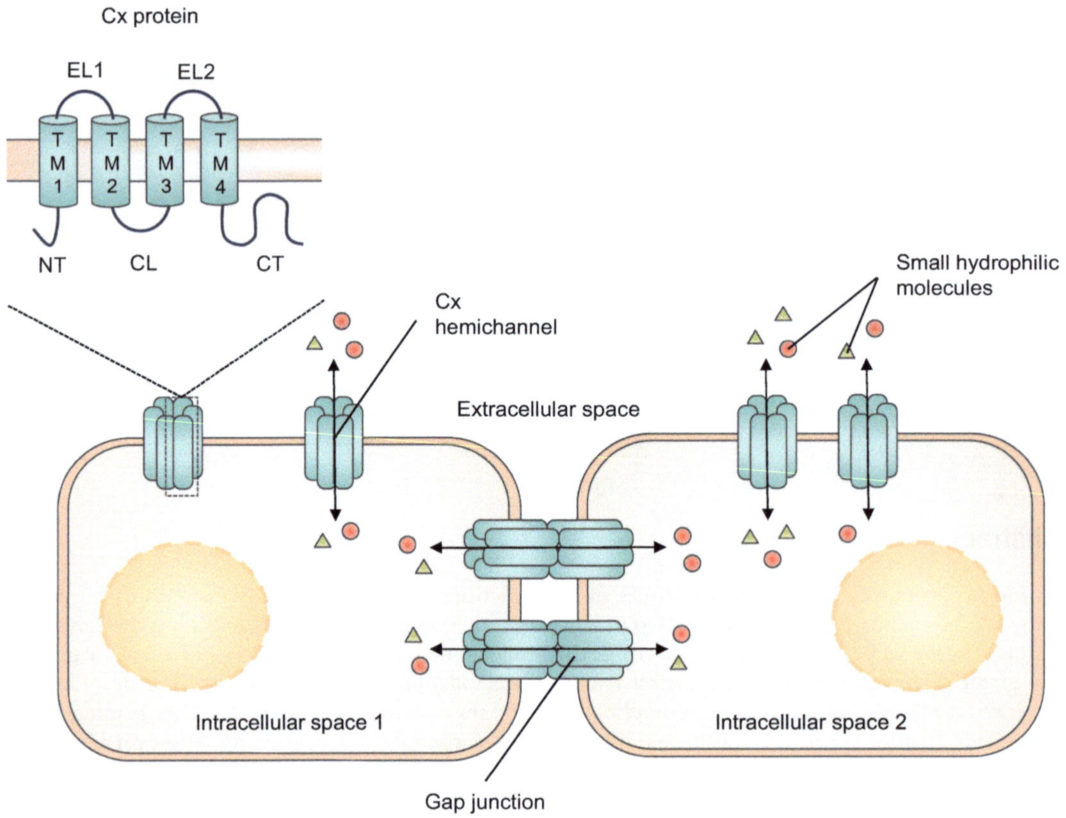

Fig. 1 Architecture of connexin (Cx) proteins, Cx hemichannels, and gap junctions. Cx proteins consist of four transmembrane regions (TM1–4), two extracellular loops (EL1-2), one cytoplasmic loop (CL), an intracellular carboxy (CT), and an amino terminus (NT). Cx hemichannels are built up by six Cx proteins. Gap junctions arise from the docking of Cx hemichannels of adjacent cells. While gap junctions mediate direct signaling between neighboring cells, open Cx hemichannels support a pathway for cellular communication between the intracellular and the extracellular space

serving as structural precursors of gap junctions, Cx hemichannels provide an autonomous pathway for cellular communication. Open Cx hemichannels allow signaling between the intracellular and the extracellular environment (Fig. 1).

Unlike gap junctions, Cx hemichannel activity is often linked to pathological conditions [7, 8]. In fact, the opening of Cx hemichannels perpetuates inflammation and cell death. Accordingly, the idea of using Cx hemichannel blockers to counteract the clinical manifestation of various inflammatory diseases is appealing [9–12]. However, robust Cx hemichannel inhibitors are lacking so far. Drug discovery in this area of research is troublesome as drug candidates often lack selectivity, specificity, potency, and/or in vivo stability [7, 12–14]. Given that Cx hemichannels and gap junctions are built up by identical molecular constituents and mediate the transfer of identical messengers, the identification of Cx hemichannel blockers that do not affect gap junctions is a challenge.

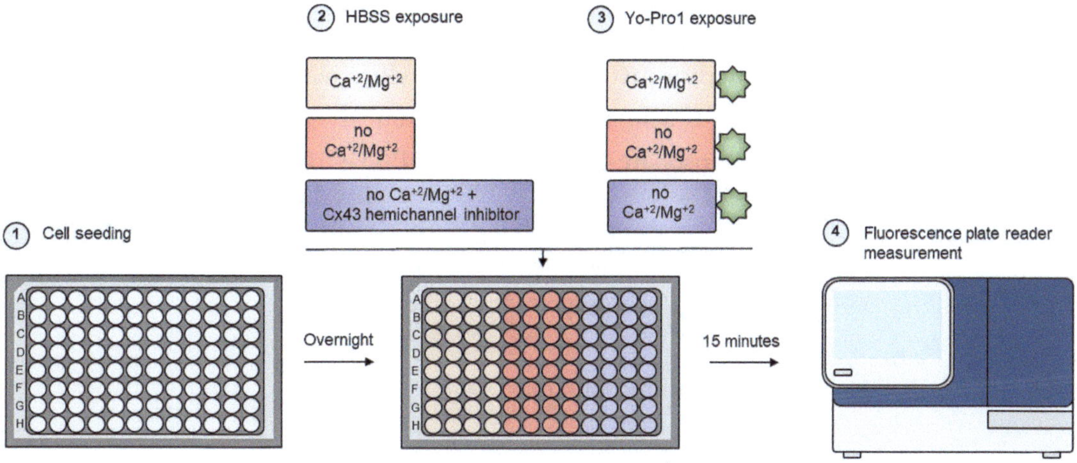

Fig. 2 Overview of Yo-Pro1 uptake assay. (1) HEK 293T cells and Cx43 overexpressing HEK 293T cells are seeded in a 96-well plate and put in an incubator overnight. (2) Add Hanks' balanced salt solution (HBSS) with calcium and magnesium, HBSS without calcium and magnesium or Cx43 hemichannel blocking solution. (3) Add Yo-Pro1 solution and place the 96-well plate in an incubator for 15 min. (4) Measure the fluorescence intensity in each well using a fluorescence plate reader

Measuring the extracellular release of biomolecules or the cytosolic uptake of fluorescent reporter dye is routinely used to indirectly assess Cx hemichannel activity [8]. In this chapter, an easy-to-use protocol for probing Cx43 hemichannel functionality is presented. Human embryonic kidney (HEK) 293T cells overexpressing Cx43 are used for this purpose [15, 16]. A calcium-free solution is used to induce Cx43 hemichannel activity [17–19]. Cx43 hemichannel opening is monitored through cytosolic uptake of Yo-Pro1, which is a cationic dye that becomes fluorescent upon binding to nucleic acids (Fig. 2). TAT-Gap19, a Cx43 mimetic peptide that specifically inhibits Cx43 hemichannel activity [20–25], is used as a positive control.

2 Materials

2.1 Generation of Cx43 Overexpressing HEK 293T Cells

Experimental work includes the use of lentiviral vectors. Follow institutional and governmental biosafety guidelines to meet required safety rules for the lentiviral transduction of Cx43. According to Belgian regulations, the generation of Cx43 overexpressing HEK 293T cells should be performed in cleanroom facilities with biosafety level-2 requirements or higher.

1. HEK 293T cells.
2. Cell culture medium: Dulbecco's modified Eagle's medium, supplemented with 10% fetal bovine serum, 0.055 g sodium pyruvate, 0.164 g L-glutamine, 50 μg/mL streptomycin sulfate, and 7.33 I.E./mL sodium benzyl penicillin.

3. Plasmids: pMD.G (VSV.G encoding plasmid), pCMVΔR8.9 (packaging plasmid), pLenti (transfer plasmid) modified to harbor mCx43 or hCx43 coding sequence.
4. Opti-Minimal Essential Medium.
5. Phosphate-buffered saline (PBS): 200 g sodium chloride, 5 g potassium chloride, 77.5 g disodium hydrogen phosphate dodecahydrate, 5 g potassium dihydrogen phosphate, pH 7.65, dissolve in 2.5 L water and autoclave.
6. 1 mg/mL polyethylenimine (PEI).
7. TrypLE Express Enzym.
8. Protamine sulfate.
9. Puromycin.
10. Cellulose acetate membrane (pore size = 0.22 μm).
11. Thermostatic water bath (37 °C).
12. CO_2 humidified incubator (37 °C, 90% humidity, 5% CO_2).
13. Class 2 biosafety cabinet.
14. Centrifuge.
15. Ultracentrifuge.
16. Flow cytometer.

2.2 Measurement of Cytosolic Yo-Pro1 Uptake

1. HEK 293T cells and Cx43 overexpressing HEK 293T cells.
2. Cell culture medium: Dulbecco's modified Eagle's medium, supplemented with 10% fetal bovine serum, 0.055 g sodium pyruvate, 0.164 g L-glutamine, 50 μg/mL streptomycin sulfate, and 7.33 I.E./mL sodium benzyl penicillin.
3. TrypLE Express Enzym.
4. Hanks' balanced salt solution (HBSS) without and with calcium and magnesium.
5. PBS: 200 g sodium chloride, 5 g potassium chloride, 77.5 g disodium hydrogen phosphate dodecahydrate, 5 g potassium dihydrogen phosphate, pH 7.65, dissolve in 2.5 L water and autoclave.
6. TAT-Gap19.
7. Yo-Pro1 iodide.
8. Thermostatic water bath (37 °C).
9. CO_2 humidified incubator (37 °C, 90% humidity, 5% CO_2).
10. Laminar flow cabinet.
11. Centrifuge.
12. Fluorescence plate reader.

3 Methods

3.1 Generation of Cx43 Expressing HEK 293T Cells

3.1.1 Lentiviral Production

1. Maintain the HEK 293T cells in cell culture medium at 37 °C with a constant supply of 5% CO_2 (*see* **Note 1**).
2. Subculture HEK 293T cells at 80–90% confluency by removing spent cell culture medium. Add 6 mL of prewarmed TrypLE Express Enzym to dissociate HEK 293T cells from a T-175 flask and incubate the cells at 37 °C for 5 min (*see* **Note 2**). Add 12 mL of prewarmed culture medium to the culture flask and distribute the medium by pipetting up and down (*see* **Note 3**).
3. Centrifuge the content of the culture flask at 125 × *g* at room temperature for 5 min. Resuspend the cell pellet with pre-warmed culture medium and subculture 1/3 of the total number of cells in a new T-175 flask (*see* **Note 4**). Incubate the HEK 293T cells in an incubator overnight.
4. Prepare the transfection mixture by adding 15 μg pMD.G, 30 μg pCMVΔR8.9, and 45 μg transfer plasmid to 5 mL of Opti-Minimal Essential Medium (*see* **Note 5**) and put 180 μg PEI to another 5 mL Opti-Minimal Essential Medium to assemble the PEI mixture. Add the PEI mixture to the transfection mixture and incubate at room temperature for 30 min (*see* **Note 6**).
5. Wash HEK 293T cells with 5 mL Opti-Minimal Essential Medium and add the transfection/PEI mixture to the cells dropwise. Shake the culture flask gently to distribute the mixture and incubate the cells in an incubator for 4 h.
6. Remove the transfection/PEI mixture and replace with 15 mL of prewarmed culture medium (*see* **Note** 7). Place the cells at 37 °C with CO_2 supply.
7. Harvest viral vectors by collecting cell supernatant 48 h after addition of the transfection/PEI mixture. Spin the collected cell supernatant at 3,000 × *g* at 4 °C for 5 min and filter through a cellulose acetate membrane (pore size = 0.22 μm) to remove cell debris (*see* **Note 8**).
8. Use ultracentrifugation to concentrate viral vectors. Spin the cell debris-free supernatant at 20,000 × *g* at 4 °C for 1.5 h. Air-dry cell pellets for 10 min and resuspend in protamine sulfate-enriched PBS solution to concentrate lentiviral vectors 1,000× (*see* **Note 9**). Pipet up and down thoroughly to resuspend lentiviral vectors. Store lentiviral vectors at −80 °C.

3.1.2 Transduction

1. Maintain the HEK 293T cells in cell culture medium at 37 °C with a constant supply of 5% CO_2 and subculture at 80–90% confluency.

Table 1
Series of dilutions of the lentiviral vectors

Dilution	Volume lentiviral vectors	Volume culture medium	Volume protamine sulfate-enriched PBS solution
1/500	4 μL of lentiviral vectors	1,996 μL	3.6 μL
1/5,000	200 μL of 1/500 dilution	1,800 μL	2.5 μL
1/10,000	750 μL of 1/5,000 dilution	750 μL	2.2 μL
1/50,000	400 μL of 1/10,000 dilution	1,600 μL	4 μL
Nontransduction	/	2,000 μL	4 μL

2. Grow 100,000 HEK 293T cells/well in 2 mL/well of cell culture medium in a six-well plate. Incubate the HEK 293T cells in an incubator overnight (*see* **Note 10**).
3. Prepare a series of dilutions of the lentiviral vectors in culture medium (Table 1) (*see* **Note 11**). Top up with protamine sulfate-enriched PBS solution (10 μg/mL) and incubate at room temperature for 15 min.
4. Remove 1 mL of the cell culture medium in the wells containing HEK 293T cells. Replace with diluted lentiviral vectors by transferring 1 mL of the 1/500, 1/5,000, 1/10,000, or 1/50,000 solution to each well. Add 1 mL of nontransduction solution to a well allocated for negative control cells. Incubate HEK 293T cells in an incubator for 72 h to evaluate transduction efficiency (*see* **Note 12**).
5. After evaluating transduction efficiency, calculate the average amount of transducing units per mL of lentiviral vectors using the following equation.

$$\text{Transducing units per mL} = \frac{\text{Number of transduced cells} \times \text{Percentage of Cx43 positive cells} \times \text{Dilution factor of lentiviral vectors}}{\text{Transduction volume in mL} \times 100}$$

6. Grow 50,000 HEK 293T cells/well in 2 mL/well of cell culture medium in a six-well plate. Incubate the HEK 293T cells in an incubator overnight.
7. Prepare 1 mL of culture medium containing 1,000,000 tranducing units (*see* **Note 13**). Top up with 2 μL of 10 μg/mL protamine sulfate-enriched PBS solution and incubate at room temperature for 15 min to prepare the transduction mixture.
8. Remove 1 mL of the cell culture medium and replace with 1 mL of the transduction mixture. Incubate HEK 293T cells in an incubator for 72 h to complete transduction.

3.1.3 Selection of Transduced Cells

1. Discard culture medium from lentivirally transduced HEK 293T cells. Add 2 mL of prewarmed cell culture medium to wells with transduced cells and incubate in an incubator overnight.
2. Prepare the selection medium by adding puromycin to the cell culture medium to a concentration of 1 μg/mL (*see* **Note 14**).
3. Replace cell culture medium with selection medium and incubate the cells in an incubator for 48 h.
4. Discard the selection medium and replace with 2 mL of prewarmed cell culture medium. Maintain the transduced HEK 293T cells in cell culture medium at 37 °C with a constant supply of 5% CO_2.
5. Subculture transduced HEK 293T cells at 80–90% confluency and gradually expand to generate stocks of Cx43 expressing HEK 293T cells (*see* **Note 15**).
6. Confirm stable transduction of HEK 293T cells at the translational level (*see* **Note 16**).

3.2 Measurement of Cytosolic Yo-Pro1 Uptake

1. Maintain HEK 293T cells and Cx43 overexpressing HEK 293T cells in cell culture medium at 37 °C with a constant supply of 5% CO_2 and subculture at 80–90% confluency.
2. Grow 10,000 HEK 293T cells/well in 0.2 mL/well of cell culture medium in a 96-well plate (*see* **Note 17**). Incubate the HEK 293T cells in an incubator overnight.
3. Prepare Cx43 hemichannel blocking solution ex tempore by diluting TAT-Gap19 in HBSS without calcium and magnesium. Add TAT-Gap19 to prepare Cx43 hemichannel blocking solutions at concentrations up to 200 μM of TAT-Gap19 (*see* **Note 18**).
4. Prepare Yo-Pro1 solution ex tempore by adding Yo-Pro1 to HBSS without and with calcium and magnesium to a concentration of 5 μM (*see* **Note 19**).
5. Discard the selection medium and replace with 0.1 mL of HBSS with calcium and magnesium, HBSS without calcium and magnesium or Cx43 hemichannel blocking solution. Immediately add 0.1 mL of the Yo-Pro1 solution to each well and incubate in an incubator for 15 min.
6. Measure the fluorescence intensity in each well using a fluorescence plate reader (*see* **Note 20**).
7. Process the results by expressing Yo-Pro1 uptake as the percentage of Yo-Pro1 relative to the uptake level triggered by HBSS without calcium and magnesium (Fig. 3) (*see* **Note 21**).

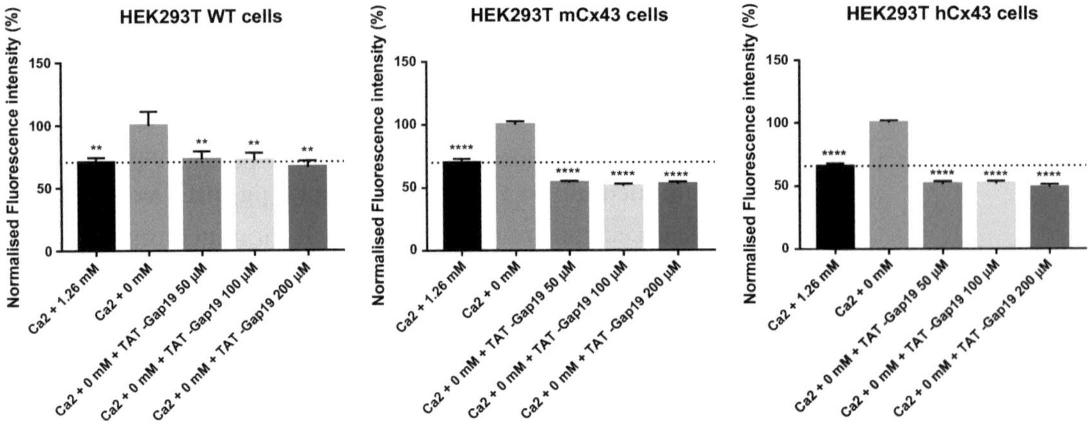

Fig. 3 Analysis of cytosolic uptake of Yo-Pro1 following transduction of HEK 293T cells. HEK 293T cells were transduced with lentiviral vectors to express mouse Cx43 (mCx43) or human Cx43 (hCx43). HEK 293T (HEK 293T WT), HEK 293T mCx43, and HEK 293T hCx43 cells were cultured. Cx43 hemichannel blocking solutions, containing TAT-Gap19 (50–200 μM), were prepared ex tempore in Hanks' balanced salt solution without calcium and magnesium. Cells were exposed to calcium (1.26 mM), calcium-free (0 mM), or Cx43 hemichannel blocking solution topped with Yo-Pro1 solution for 15 min. Cytosolic uptake of Yo-Pro1 was measured and expressed as the percentage of Yo-Pro1 relative to the uptake level triggered by calcium-free solution ($n = 3$, $N = 4$). Results were analyzed by parametric one-way analysis of variance followed by post hoc tests with Dunnett's correction. Data were expressed as means ± standard error of the mean ($p \leq 0.01$; $p \leq 0.0001$)

4 Notes

1. To maintain cultured HEK 293T cells, the subculture procedure should be performed every 3 days. By passaging frequently, the cells remain individual units, which supports high transfection efficiency.
2. TrypLE Express Enzym is a dissociating agent that cleaves peptide bonds to allow dislocation of HEK 293T cells from plasticware. A light microscope is often used to confirm complete detachment. Slightly tapping the culture flask or extending the incubation time with TrypLE stimulates cell detachment.
3. Use of TrypLE Express Enzym bypasses the need for trypsin inhibitors. Dilution with cell culture medium is sufficient to stop the detachment of the cells.
4. The ratio of subcultures is flexible because the cell growth rate can vary. The goal is to bring HEK 293T cells at 70–80% confluency.
5. Ethanol-precipitated plasmids are used for transfection. After centrifugation with 70% ethanol at 14,000 × g for 10 min, air-dry pellets are dissolved in endotoxin-free water to a concentration of 1 μg/mL.

6. PEI is a synthetic polycation that condenses plasmids into nanoparticles. It is recommended to pipet up and down the PEI mixture in the transfection mixture. Avoid vortexing for the preparation of PEI/DNA complexes.
7. Removal of the transfection/PEI mixture is important because the transfection time affects cell viability.
8. The collected cell supernatant is kept at 4 °C for short-term storage. Preserve at −80 °C for long-term storage.
9. PBS is supplemented with protamine sulfate to a concentration of 10 μg/mL. The addition of polycations improves the efficiency of lentiviral transduction. It facilitates the cellular adherence of lentiviral vectors.
10. Add culture media to unused wells to limit evaporation from cell culture plates.
11. Ice is recommended for thawing the prepared lentiviral vectors. Lentiviral vectors should be kept on ice while performing transduction. Avoid multiple freeze-thaw cycles by making aliquots.
12. After titrating lentiviral vectors, examination of transgene Cx43 expression is used to evaluate transduction efficiency. Flow cytometry analysis is a simple method for selecting appropriate volumes of lentiviral vectors. Staining tranduced and nontransduced HEK 293T cells with fluorescent-labeled Cx43 antibodies is a straightforward approach to evaluate the result of transduction.
13. For the transduction of HEK 293T cells at a multiplicity of infection of 10, a volume of lentiviral vectors containing 1,000,000 tranducing units is used.
14. Transduced cell populations should be enriched upon transduction. The pLenti plasmid used is specifically designed for puromycin-mediated selection. A working concentration of 1 μg/mL is recommended to remove nontransduced HEK 293T cells. The optimal puromycin concentration varies between cell types. This value is determined by evaluating dose-dependent effects of puromycin on the viability of HEK 293T cells. A working concentration of puromycin is the minimum antibiotic concentration required to kill all nontransduced cells within 48 h.
15. Transduced HEK 293T cells are cryopreserved in cell culture medium containing 10% dimethylsulfoxide. Aliquots of cell suspensions, containing 4,000,000 cells/mL, are placed in cryogenic storage vials and frozen slowly at 1 °C/min to −80 °C using a programmable cooler before being transferred to liquid nitrogen.

16. To indicate an overexpression of Cx43 proteins after transduction, immunoblot, immunocytochemistry, flow cytometry, or enzyme-linked immunosorbent assays should be performed.
17. For cell-based fluorescent assays, clear-bottom black plates are used. This reduces background signal and signal crosstalk between wells.
18. Working concentrations up to 200 μM TAT-Gap19 are used. This concentration of TAT-Gap19 does not induce cytotoxicity in HEK 293T cells and Cx43 overexpressing HEK 293T cells as shown by a lactate dehydrogenase assay.
19. Yo-Pro1 is a monomeric cyanine nucleic acid stain and must be stored in the dark. Avoid photobleaching when performing experimental procedures with Yo-Pro1.
20. To measure the fluorescence intensity, preheat the plate reader at 37 °C and set up a protocol to read wells from the bottom with an excitation wavelength range of 491–509 nm.
21. Statistically process the results using an appropriate test. When comparing three or more groups, results can be evaluated by one-way analysis of variance followed by post hoc tests.

Acknowledgments

This work was financially supported by the European Research Council (Proof-of-Concept grant 861913), the European Future and Emerging Technologies program (grant 858014), the Research Foundation Flanders-Belgium (grants G012318N, G020018N and G0F7219N), the Methusalem program of the Flemish Government and the University Hospital of the Vrije Universiteit Brussel-Belgium (Scientific Fund Willy Gepts). The authors wish to thank Fien Haenen and Arne Loosen for their dedicated technical assistance.

References

1. Lukowicz-Bedford R, Farnsworth DR, Miller AC (2021) Connexinplexity: the spatial and temporal expression of connexin genes during vertebrate organogenesis. bioRxiv preprint
2. Oyamada M, Oyamada Y, Takamatsu T (2005) Regulation of connexin expression. Biochim Biophys Acta Biomembr 1719:6–23
3. Beyer EC, Paul DL, Goodenough DA (1987) Connexin43: a protein from rat heart homologous to a gap junction protein from liver. J Cell Biol 105:2621–2629
4. Totland MZ, Rasmussen NL, Knudsen LM et al (2020) Regulation of gap junction intercellular communication by connexin ubiquitination: physiological and pathophysiological implications. Cell Mol Life Sci 77:573–591
5. Zhu Y (2022) Gap junction-dependent and -independent functions of Connexin43 in biology. Biology (Basel) 11:1–19
6. Liu W, Cui Y, Wei J et al (2020) Gap junction-mediated cell-to-cell communication in oral development and oral diseases: a concise review of research progress. Int J Oral Sci 12:1–9
7. Van Campenhout R, Gomes AR, De Groof TWM et al (2021) Mechanisms underlying

connexin hemichannel activation in disease. Int J Mol Sci 22:1–14

8. Peng B, Xu C, Wang S et al (2022) The role of Connexin Hemichannels in inflammatory diseases. Biology (Basel) 11:1–20
9. Cooreman A, Van Campenhout R, Ballet S et al (2019) Connexin and Pannexin (hemi)-channels: emerging targets in the treatment of liver disease. Hepatology 69:1317–1323
10. Delvaeye T, De Smet MAJ, Verwaerde S et al (2019) Blocking connexin43 hemichannels protects mice against tumour necrosis factor-induced inflammatory shock. Sci Rep 9:1–12
11. Mugisho OO, Green CR, Zhang J et al (2019) Connexin43 hemichannels: a potential drug target for the treatment of diabetic retinopathy. Drug Discov Today 24:1627–1636
12. Retamal MA, Fernandez-Olivares A, Stehberg J (2021) Over-activated hemichannels: a possible therapeutic target for human diseases. Biochim Biophys Acta Mol basis Dis 1867:1–15
13. Caufriez A, Böck D, Martin C et al (2020) Peptide-based targeting of connexins and pannexins for therapeutic purposes. Expert Opin Drug Discov 15:1213–1222
14. Willebrords J, Maes M, Crespo Yanguas S et al (2017) Inhibitors of connexin and pannexin channels as potential therapeutics. Pharmacol Ther 180:144–160
15. Goyvaerts C, De Groeve K, Dingemans J et al (2012) Development of the Nanobody display technology to target lentiviral vectors to antigen-presenting cells. Gene Ther 19:1133–1140
16. Breckpot K, Dullaers M, Bonehill A et al (2003) Lentivirally transduced dendritic cells as a tool for cancer immunotherapy. J Gene Med 5:654–667
17. Lopez W, Ramachandran J, Alsamarah A et al (2016) Mechanism of gating by calcium in connexin hemichannels. Proc Natl Acad Sci USA 113:7986–7995
18. Thimm J, Mechler A, Lin H et al (2005) Calcium-dependent open/closed conformations and interfacial energy maps of reconstituted hemichannels. J Biol Chem 280:10646–10654
19. Harris AL, Contreras JE (2014) Motifs in the permeation pathway of connexin channels mediate voltage and Ca2+ sensing. Front Physiol 5:1–10
20. Maes M, Crespo Yanguas S, Willebrords J et al (2017) Connexin hemichannel inhibition reduces acetaminophen-induced liver injury in mice. Toxicol Lett 278:30–37
21. Crespo Yanguas S, da Silva TC, Pereira IVA et al (2018) TAT-Gap19 and Carbenoxolone alleviate liver fibrosis in mice. Int J Mol Sci 19:1–17
22. Walrave L, Pierre A, Albertini G et al (2018) Inhibition of astroglial connexin43 hemichannels with TAT-Gap19 exerts anticonvulsant effects in rodents. Glia 66:1788–1804
23. Walrave L, Vinken M, Albertini G et al (2016) Inhibition of connexin43 hemichannels impairs spatial short-term memory without affecting spatial working memory. Front Cell Neurosci 10:1–10
24. Willebrords J, Cogliati B, Pereira IVA et al (2017) Inhibition of connexin hemichannels alleviates non-alcoholic steatohepatitis in mice. Sci Rep 7:1–11
25. Wang N, De Vuyst E, Ponsaerts R et al (2013) Selective inhibition of Cx43 hemichannels by Gap19 and its impact on myocardial ischemia/reperfusion injury. Basic Res Cardiol 108:1–16

Chapter 7

A Dye Uptake Assay to Measure Large-Pore Channel Activity in *Trypanosoma cruzi*

José Luis Vega, Aníbal García, Jorge González, and Juan C. Sáez

Abstract

Large-pore channels allow the exchange of ions and molecules between the intra- and extracellular compartments. These channels are structures formed by several protein families with little or no evolutionary linkages that include connexins (Cxs), pannexins (Panxs), innexins (Inxs), CALHM1, and LRRC8 proteins. Recently, we have described the unnexins (Unxs) proteins expressed in *Trypanosoma cruzi* (*T. cruzi*) that also is like to form large-pore channels at the plasma membrane. In this chapter, we describe a dye uptake method for evaluating the unnexin-formed channel function in *T. cruzi*, as well as the methods for evaluating their participation in the transformation of trypomastigotes into amastigotes. These methods can facilitate understanding the role of large-pore channels in the parasite's biology.

Key words Unnexin, Chagas diseases, Parasitic diseases

1 Introduction

Large-pore channels are structures formed by several protein families with no evolutionary linkages, including Cxs, Panxs, Inxs, Unxs, CALHM1, and LRRC8 proteins [1, 2]. Despite little sequence homology, the large-pore channel members have similar topology with four transmembrane helices arranged similarly [1, 2]. Recent evidence suggests that pathogenic protozoan parasites such as *T. cruzi* have large-pore channels formed by protein homologs of innexins, named unnexins, which could play a role in infection [2]. Here, we present a protocol for dye uptake and transformation assay. These protocols will be helpful for future experiments in the field of large-pore channels study in parasitic diseases.

Fabio Mammano and Mauricio Retamal (eds.), *Connexin Hemichannels: Methods and Protocols*, Methods in Molecular Biology, vol. 2801, https://doi.org/10.1007/978-1-0716-3842-2_7,

2 Materials

2.1 Parasites Culture

1. *T. cruzi* H510 C8C3 clone. This clone was isolated from a domiciliary *Triatoma dimidiata* found in an area endemic for Chagas disease in Costa Rica [3] (*see* **Note 1**).
2. ddH_2O.
3. 1 mL of 0.5% Trypsin-EDTA (10X) no phenol red.
4. Liver infusion tryptose (LIT) medium: 7.4 g tryptose, 5.1 g liver infusion broth, 2.7 g NaCl, 0.4 g KCl, 14.5 g Na_2HPO_4, supplemented with 0.25 g/L bovine hemin, 10% (v/v) of glucose solution (sterile), and 10% (v/v) fetal bovine serum (FBS) heat-inactivated complement system.
5. Phosphate buffer saline (PBS), pH 7.4.
6. Sterile disposable glass serological pipets, 5 mL.
7. Sterile disposable glass serological pipets, 10 mL.
8. Disposable glass Pasteur pipets.
9. 2.0 mL microcentrifuge tube.
10. Cryogenic vials.
11. T25 flask with vent cap.
12. T75 flask with vent cap.
13. 15 mL centrifuge tubes.
14. 50 mL centrifuge tubes.
15. PVC-coated 1 L round media storage bottles, with GL45 screw cap.
16. PVC-coated 500 mL round media storage bottles, with GL45 screw cap.
17. PVC-coated 100 mL round media storage bottles, with GL45 screw cap.
18. Motorized pipet controller for use with plastic serological pipets.
19. Neubauer improved hemocytometer counting chamber, depth 0.1 mm.
20. Refrigerator 4 °C.
21. Temperature controlled water bath.
22. Vacuum system.
23. Biological safety cabinet with laminar flow and UV light.
24. Light-inverted microscope dedicated for *T. cruzi* culture.
25. Centrifuge.
26. 26 °C incubator dedicated for *T. cruzi*.

27. CO_2 incubator (37 °C, 5% CO_2, 95% humidity) dedicated for *T. cruzi*.

2.2 Cell Line Culture

1. Vero cells (ATCC, cat. no. CCL-81) (*see* **Note 2**).
2. CO_2 incubator (37 °C, 5% CO_2, 95% humidity) dedicated for *T. cruzi*.
3. RPMI 1640 medium.
4. FBS heat-inactivated complement system.
5. PBS pH 7.4.
6. Ethyl alcohol 70° (*see* **Note 3**).
7. 0.5% Trypsin-EDTA (10X), no phenol red.
8. Penicillin–streptomycin solution with penicillin 10.000 units/mL and streptomycin 10 mg/mL.
9. Cryogenic vials.
10. 0.4% (wt/vol) Trypan blue solution.
11. Tissue culture plate, 35 mm diameter, with treated surface.
12. Tissue culture plate, 60 mm diameter, with treated surface.
13. Sterile disposable glass serological pipets, 5 mL.
14. Sterile disposable glass serological pipets, 10 mL.
15. T25 flask with vent cap.
16. T75 flask with vent cap.
17. 15 mL centrifuge tubes.
18. 50 mL centrifuge tubes.
19. Neubauer improved hemocytometer counting chamber, depth 0.1 mm.
20. Motorized pipet controller for use with plastic serological pipets.
21. General purpose water baths.
22. Refrigerator 4 °C.
23. Biological safety cabinet with laminar flow and UV light.
24. Light-inverted microscope dedicated for *T. cruzi* culture.
25. Centrifuge.
26. CO_2 incubator (37 °C, 5% CO_2, 95% humidity).

2.3 Dye Uptake Assay

1. PBS pH 7.4.
2. Dye uptake solution: 154 mM NaCl, 5.4 mM KCl, 2.3 mM $CaCl_2$, 2.1 mM $MgCl_2$, 5 mM glucose, 10 mM HEPES, and 5 μM dye. Prepare a fresh solution before use.

3. DCFS solution: 154 mM NaCl, 5.4 mM KCl, 0.5 mM ethylenediaminetetraacetic acid (EDTA), 5 mM glucose, 10 mM HEPES, and 5 μM dye. Prepare a fresh solution before use.
4. 10 mM (100×) carbenoxolone solution. Prepare a fresh solution before use.
5. 40 mM (100×) probenecid solution. Prepare a fresh solution before use.
6. 4 mM (100×) 18β-glycyrrhetinic acid solution. Prepare a fresh solution before use.
7. 20 mM (100×) lanthanum chloride solution. Prepare a fresh solution before use.
8. 10 mg/mL ethidium bromide stock solution (*see* **Note 4**).
9. 5 mg/mL 4′,6-diamidino-2-phenylindole (DAPI) stock solution (*see* **Note 5**).
10. 1 mM YO-PRO-1 iodide stock solution.
11. 25 mg/mL dextran, fluorescein 70.000 MW stock solution.
12. ProLong Gold Antifade reagent.
13. Methanol (≥99.95%, HPLC) (*see* **Note 6**).
14. 4% (wt/vol) paraformaldehyde. Prepare a fresh solution before use (*see* **Note 7**).
15. Microscope slides 75 × 25 mm.
16. Eppendorf Research Plus single-channel 20–200 μL pipette.
17. Eppendorf Research Plus single-channel 100–1.000 μL pipette.
18. Sterile pipette tips.
19. 12-well plates.
20. Leica TCS SP8 confocal laser-scanning microscope.
21. Epifluorescence microscope.

2.4 Transformation Assay

1. Transformation medium: DMEM buffered with 20 mM 2-(N-morpholino) ethanesulfonic acid, pH 5.0.
2. Methanol (≥99.95%, HPLC grade) (*see* **Note 8**).
3. 10 mM (100×) carbenoxolone solution. Prepare a fresh solution before use.
4. 40 mM (100×) probenecid solution. Prepare a fresh solution before use.
5. 4 mM (100×) 18ß-glycyrrhetinic acid solution. Prepare a fresh solution before use.
6. 20 mM (100×) lanthanum chloride solution. Prepare a fresh solution before use.
7. Microscope slides 75 × 25 mm.

8. Eppendorf Research Plus single-channel 20–200 μL pipette.
9. Eppendorf Research Plus single-channel 100–1.000 μL pipette.
10. Sterile pipette tips.
11. Sterile disposable glass serological pipets, 5 mL.
12. Sterile disposable glass serological pipets, 10 mL.
13. Motorized pipet controller for use with plastic serological pipets.
14. Neubauer-improved hemocytometer counting chamber, depth 0.1 mm.
15. CO_2 incubator (37 °C, 5% CO_2, 95% humidity).
16. 12-well plates.
17. Light-inverted microscope dedicated for *T. cruzi* culture.

3 Methods

3.1 T. cruzi Trypomastigotes Culture

1. *T. cruzi* trypomastigotes H510 C8C3 clone are maintained in cell cultures of Vero cells grown RPMI 1640 supplemented with 10% FBS at 37 °C in a 5% CO_2 incubator (*see* **Note 9**).
2. To prepare Vero cells for infection. Move the cryogenic vial containing Vero cells from liquid nitrogen.
3. Transfer the cells to a 15 mL Falcon tube containing 9 mL of prewarmed RPMI 1640 medium and centrifuge at 2.000×*g* for 5 min at room temperature. Then, discard the supernatant and resuspend the pellet in 5 mL of RPMI 1640 medium with 10% FBS.
4. Keep the cells at 37 °C in a 5% CO_2 incubator.
5. After 6–8 h of attachment, replace the RPMI 1640-containing medium with 5 mL RPMI 1640 medium with 10% FBS.
6. Culture the cells at 37 °C for 2–3 days until ~80% confluence.
7. Attach an autoclaved borosilicate glass Pasteur pipette to a vacuum line and aspirate the growth medium from the dish.
8. Gently rinse the cell monolayer with 1–2 mL of sterile PBS. Use the Pasteur pipette to aspirate PBS between washes.
9. Add 1 mL of 0.5% Trypsin-EDTA (10X) and place the dish in the cells at 37 °C in a 5% CO_2 incubator for 3 min.
10. Remove the dish from the incubator and check whether cells have detached using a light-transmitted microscope. Once cells are detached, add 1 mL of complete RMPI 1640 to inhibit the trypsin.
11. Remove cells using a 1.000 mL pipette and place the cell suspension in a 2 mL Eppendorf tube.

12. Centrifuge the tube at 2.000×*g* for 5 min at room temperature.
13. Remove supernatant and add 1 mL of complete RPMI 1640. Gently suspend the cell pellet by pipetting up and down to break apart any cell aggregates.
14. Count the cell number using a Neubauer chamber.
15. Calculate the volume required for seeding 1×10^4 cells and add this volume to a new T75 culture flask.
16. Add 4 mL of the complete RPMI 1640 and keep the cells at 37 °C in a 5% CO_2 incubator.
17. Change medium every day and wait until approximately 80% confluence.
18. To prepare *T. cruzi* trypomastigotes for infection. Move the cryogenic vial containing *T. cruzi* trypomastigotes from liquid nitrogen.
19. Transfer the parasites to a centrifuge tube containing 5 mL of prewarmed RPMI 1640 medium.
20. Centrifuge the tube at 2.000×*g* for 5 min at room temperature.
21. Remove the supernatant and resuspend the pellet in 5 mL of prewarmed RPMI 1640 medium.
22. Count the cell number using a Neubauer chamber.
23. Add trypomastigotes in a 4:1 ratio and keep the flask at 37 °C in a 5% CO_2 incubator.
24. To visualize the cells every day and change the medium. After 5 days, the parasites begin to leave the cells.
25. Collect the motile extracellular trypomastigotes from infected Vero supernatants into a 15 mL tube.
26. Spin the parasites at 2.000×*g* for 15 min at 4 °C.
27. Prepare the 1×10^6 parasites/mL for experiment.

3.2 T. cruzi *Epimastigotes Culture*

1. Move the cryogenic vial containing *T. cruzi* epimastigotes from liquid nitrogen.
2. Transfer the parasites to a Falcon tube containing 5 mL of LIT medium.
3. Centrifuge the tube at 2.000×*g* for 5 min at room temperature.
4. Remove the supernatant and resuspend the pellet in 5 mL of LIT medium and transfer to T25 flask.
5. Culture the epimastigotes at 26 °C and use in logarithmic phase of growth (3–6 day).

3.3 Dye Uptake Assay in T. cruzi

1. Clean all working surfaces with 70% (vol/vol) ethanol before commencing any procedure.

Table 1
Characteristics of fluorescent dyes commonly used for dye uptake assays

Dye	Molecular weight (Da)	Charge	Excitation/emission (nm)	References
YO-PRO-1	629.3	+2	491/509	[2]
Evans blue	960.8	−4	545/595	[4]
Ethidium bromide	394.0	+1	518/605	[5]
4′-6-diamidino-2-phenylindole (DAPI)	350.7	+2	358/461	[2]
Fluorescein isothiocyanate dextran (FITC-dextran)	70.000	0	491/516	[2]

2. Wash *T. cruzi* trypomastigotes or epimastigotes three times with 1× PBS (2.000×*g* for 10 min).
3. Count the parasites number using a Neubauer chamber.
4. Calculate the volume required for 1 × 10^6 parasites/mL and place the parasites in their respective wells of 12-well plates.
5. Add the necessary volume to achieve a final concentration of 100 μM carbenoxolone, 400 μM probenecid, 40 μM 18-β-glycyrrhetinic acid, or 200 μM lanthanum chloride blockers, and incubate the parasites for 30 min at 26 °C (epimastigotes) or 37 °C (trypomastigotes) in a 5% CO_2 incubator.
6. Add 500 μL of dye solution or DCFS dye-containing solution, freshly prepared, and incubate for 30 min with parasites in a 5% CO_2 incubator. Evaluate different dyes (*see* Table 1).
7. Add 500 μL of 4% (wt/vol) solution of paraformaldehyde, freshly prepared, and incubate for 5 min.
8. Wash *T. cruzi* trypomastigotes or epimastigotes three times with 1× PBS (2.000×*g* for 10 min).
9. Mounted in ProLong Gold Antifade reagent on a glass slide.
10. Acquire fluorescence images using an epifluorescence microscope.
11. Analyze the fluorescence intensity of regions of interest placed on the cells using the Image J software and express them as arbitrary units. Subtract the background value for each of regions of interest.

3.4 Transformation in Axenic Medium of Trypomastigotes into Amastigotes-like Organisms

1. Obtain the motile extracellular *T. cruzi* trypomastigotes from infected Vero supernatants.
2. Wash *T. cruzi* trypomastigotes three times with 1× PBS (2.000×*g* for 10 min).

3. Count the parasites number using a Neubauer chamber.
4. Calculate the volume required for 2×10^7 parasites/mL and place the parasites in their respective wells of 12-well plates.
5. Add the necessary volume to achieve a final concentration of 100 μM carbenoxolone, 400 μM probenecid, 40 μM 18-β-glycyrrhetinic acid, or 200 μM lanthanum chloride blockers, and incubate the trypomastigotes for 30 min at 37 °C in a 5% CO_2 incubator.
6. Add 500 μL of transformation medium and incubate trypomastigote for 4 h at 37 °C in a 5% CO_2 incubator.
7. To determine a percentage of transformed parasites by microscopically scoring 500 cells in each well in a blinded fashion.

4 Notes

1. *T. cruzi* is a causal agent of Chagas diseases and must be handled in accordance with both government and institutional guidelines and regulations. Personal protective equipment (PPE) consisting of a laboratory coat, eye or fascial protection, and gloves should be worn during all procedures. Avoid contact of media containing *T. cruzi* parasites with open wounds. All work must be performed within a microbiological safety cabinet class II, and waste should be discarded by appropriate methods dictated within each establishment.
2. The cell lines should be regularly checked to ensure that they are not infected with mycoplasma.
3. Ethanol is flammable. Handle it with appropriate PPE and care.
4. Ethidium bromide is a known mutagen. Handle with gloves and dispose of safely in accordance with institutional regulations.
5. DAPI is a known mutagen. Handle with gloves and dispose of safely in accordance with institutional regulations.
6. Methanol is toxic and very flammable. Handle the compound with appropriate PPE and care. Avoid inhalation. Using a hazardous fume hood is recommended.
7. It is harmful if inhaled, use adequate ventilation.
8. Methanol is toxic and very flammable. Handle the compound with appropriate PPE and care. Avoid inhalation. Using a hazardous fume hood is recommended.
9. All procedures involving live organisms must conform to ethics and biosafety Institutional regulations.

References

1. Syrjanen J, Michalski K, Kawate T, Furukawa H (2021) On the molecular nature of large-pore channels. J Mol Biol 433:166994
2. Güiza J, García A, Arriagada J, Gutiérrez C, González J, Marquez-Miranda V, Alegría-Arcos M, Duarte Y, Rojas M, González-Nilo F, Sáez JC, Vega JL (2022) Unnexins: homologs of innexin proteins in *Trypanosomatidae* parasites. J Cell Physiol 237:1547–1560
3. San Francisco J, Astudillo C, Vega JL, Catalán A, Gutiérrez B, Araya JE, Zailberger A, Marina A, García C, Sánchez N, Osuna A, Vilchez S, Ramírez MI, Macedo J, Feijoli VS, Palmisano G, González J (2022) *Trypanosoma cruzi* pathogenicity involves virulence factor expression and upregulation of bioenergetic and biosynthetic pathways. Virulence 13:1827–1848
4. Cea LA, Cisterna BA, Puebla C, Frank M, Figueroa XF, Cardozo C, Willecke K, Latorre R, Sáez JC (2013) De novo expression of connexin hemichannels in denervated fast skeletal muscles leads to atrophy. Proc Natl Acad Sci USA 110: 16229–16234
5. Retamal MA, Froger N, Palacios-Prado N, Ezan P, Sáez PJ, Sáez JC, Giaume C (2007) Cx43 hemichannels and gap junction channels in astrocytes are regulated oppositely by proinflammatory cytokines released from activated microglia. J Neurosci 27:13781–13792

Chapter 8

Measurement of Ca^{2+} Uptake Through Connexin Hemichannels

Chiara Nardin and Fabio Mammano

Abstract

Increasing evidence points to deregulated flux of ionized calcium (Ca^{2+}) mediated by hyperactive mutant connexin (Cx) hemichannels (HCs) as a common gain-of-function etiopathogenetic mechanism for several diseases, ranging from skin disorders to nervous system defects. Furthermore, the opening of nonmutated Cx HCs is associated with an impressive list of widespread diseases including, but not limited to, ischemia/stroke, Alzheimer's disease, and epilepsy. HC inhibitors are attracting a growing attention due to their therapeutic potential for numerous pathologies. This chapter describes a quantitative method to measure Ca^{2+} uptake though HCs expressed in cultured cells. The assay we developed can be used to probe HC activity as wells as to test HC inhibitors. Furthermore, with minor changes it can be easily adapted to high-throughput high-content platforms and/or primary cells and microtissues.

Key words Hemichannel functionality, HaCaT cells, Connexin43, TAT-Gap19, GCaMP6s, Ionomycin, Lentiviral transduction

1 Introduction

Cx HCs can dock head-to-head in the extracellular space to form intercellular gap junction channels (IGJCs) [1–4], whereas unpaired HCs in the cell plasma membrane may open to exchange inorganic anions and cations, as well as larger metabolites and signaling molecules with molecular weight below 1–2 kDa between the cytoplasm and the extracellular milieu [5].

Increasing evidence points to leaky or hyperactive HCs as a common gain-of-function etiopathogenetic mechanism for monogenic diseases affecting the facial appearance, dentition (small and carious teeth), eyes (microphthalmia, microcornea), fingers (syndactyly), inner ear (deafness), skin (various disorders), lens (cataracts), nervous system (myelination defects), and heart (atrial fibrillation); reviewed in [6, 7]. Ca^{2+} is both a natural permeant [8–12] and a key regulator of HC gating [13–15] (reviewed in [16–

Fabio Mammano and Mauricio Retamal (eds.), *Connexin Hemichannels: Methods and Protocols*, Methods in Molecular Biology, vol. 2801, https://doi.org/10.1007/978-1-0716-3842-2_8,

21]), and deregulated Ca^{2+} flux through HCs has been implicated in a number of disease-linked Cx mutations, including those affecting the skin [22]. In addition, the opening of wild type (*wt*, i.e., nonmutated) HCs is associated with an impressive list of widespread diseases that affect a major proportion of the world population, including ischemia/stroke, Alzheimer's disease, epilepsy, liver fibrosis and cirrhosis, nonalcoholic steatohepatitis, inflammation, and others [23, 24]. HC inhibitors are attracting a growing attention as potential therapeutics [25–27], including antibodies targeting the extracellular domain of HCs [28–32].

HaCaT cells are a widely used spontaneously immortalized line of aneuploid human keratinocytes [33] known to express different Cx isoforms, among which the prevalent one is Cx43 [34, 35]. HaCaT cells have been mostly utilized in the context of IGJC-mediated cell–cell communication [36, 37]. However, also HC activity has been recently observed in ATP release assays performed in this cell line, either parental HaCaT cells [38] or cells overexpressing Cx26 or Cx30 [29, 31]. Other classical methods to probe HC function are dye uptake or leakage assays, based on passive diffusion of fluorescent molecules across open HCs in the cell plasma membrane [39]. Assays of this type have been mainly used in model cell systems, including HaCaT cells [29], but occasionally also in native tissues [40] or in vivo [41].

This chapter provides detailed information to quantify HC-mediated Ca^{2+} uptake, which can be used to probe HC activity as well as to evaluate the efficacy of HC inhibitors [42]. To develop the Ca^{2+} uptake assay, we exploited the well-known fact that lowering the extracellular Ca^{2+} concentration ($[Ca^{2+}]_{ex}$) increases the open probability of HCs formed by several types of Cxs [13–21]. To track changes of the cytosolic Ca^{2+} concentration ($[Ca^{2+}]_{cyt}$) with high sensitivity, we took advantage of GCaMP6s, a Ca^{2+}-selective genetically encoded indicator [43]. We produced a lentivirus (LV) encoding for GCaMP6s utilizing a standard protocol [44] and used it to infect HaCaT cells, generating GCaMP6s-expressing cells (HaCaT-GCaMP6s, *see* Fig. 1); however, the assay can be conducted in virtually any type of cell with minor adjustments.

We performed experiments on confluent HaCaT-GCaMP6s cell cultures using a custom-built spinning disk confocal fluoresce microscope [45] equipped with a 20× water-immersion objective. The wide field of view (FOV) achieved with this objective (740 × 630 μm) allowed us to collect GCaMP6s fluorescence emission from more than 100 cells per culture. The Ca^{2+} uptake assay consisted in imaging $[Ca^{2+}]_{cyt}$ variation reported by GCaMP6s fluorescence intensity (F) changes (ΔF) in cells that were previously kept in a low Ca^{2+}-content extracellular solution and then abruptly exposed to a high $[Ca^{2+}]_{ex}$ (*see* Fig. 2).

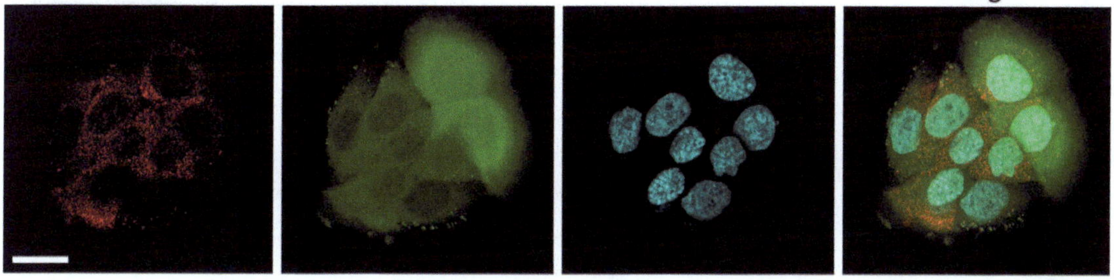

Fig. 1 Expression of native Cx43 in HaCaT-GCaMP6s cells. Immunofluorescence staining with an anti-Cx43 antibody is shown in red; endogenous GCaMP6s signal is shown in green; nuclei stained with the nuclear staining dye 4′,6-diamidino-2-phenylindole (DAPI) are shown in cyan. Scale bar: 20 μm

To confirm that the observed Ca^{2+} influx was mediated by HCs, we used flufenamic acid (FFA), well-known aspecific blocker, as well as TAT-Gap19, the cell-penetrating moiety of a mimetic peptide that inhibits Cx43 HCs without affecting IGJCs [46]. For negative controls, we used abEC1.1m, a monoclonal antibody formatted as a single-chain fragment variable (scFv)-fragment constant (Fc) diabody [29] (with human scFv and mouse Fc domain) that specifically binds to and inhibits Cx26, Cx30, and Cx32 HCs, but is not effective on Cx43 HCs [30]. We also used abEC1.1h-IgG, a variant of the same antibody formatted as a fully human (h) immunoglobulin G (IgG).

2 Materials

2.1 Generation of Stable HaCaT-GCaMP6s Cell Line

All solutions mentioned in this section are sterile-filtered and all reagents and plastic materials are suitable for cell culture.

Precaution Experimental work includes the use of a LV vector. Follow institutional and governmental biosafety guidelines to meet required safety rules for the LV transduction of GCaMP6s. The generation of GCaMP6s-expressing cells must be performed in cleanroom facilities with biosafety level-2 (BSL-2) requirements or higher.

2.1.1 LV Production

1. HEK 293T cells, passage number <20 (*see* **Note 1**).
2. Lentiviral vector plasmid: pHAGE-RSV-GCaMP6s (Plasmid #80146, Addgene; http://n2t.net/addgene:80146; RRID: Addgene_80146).
3. Second-generation lentiviral packaging vectors (pMD2.G and psPAX2, Plasmids #12259 and #12260, Addgene, respectively; https://www.addgene.org/browse/article/836/).

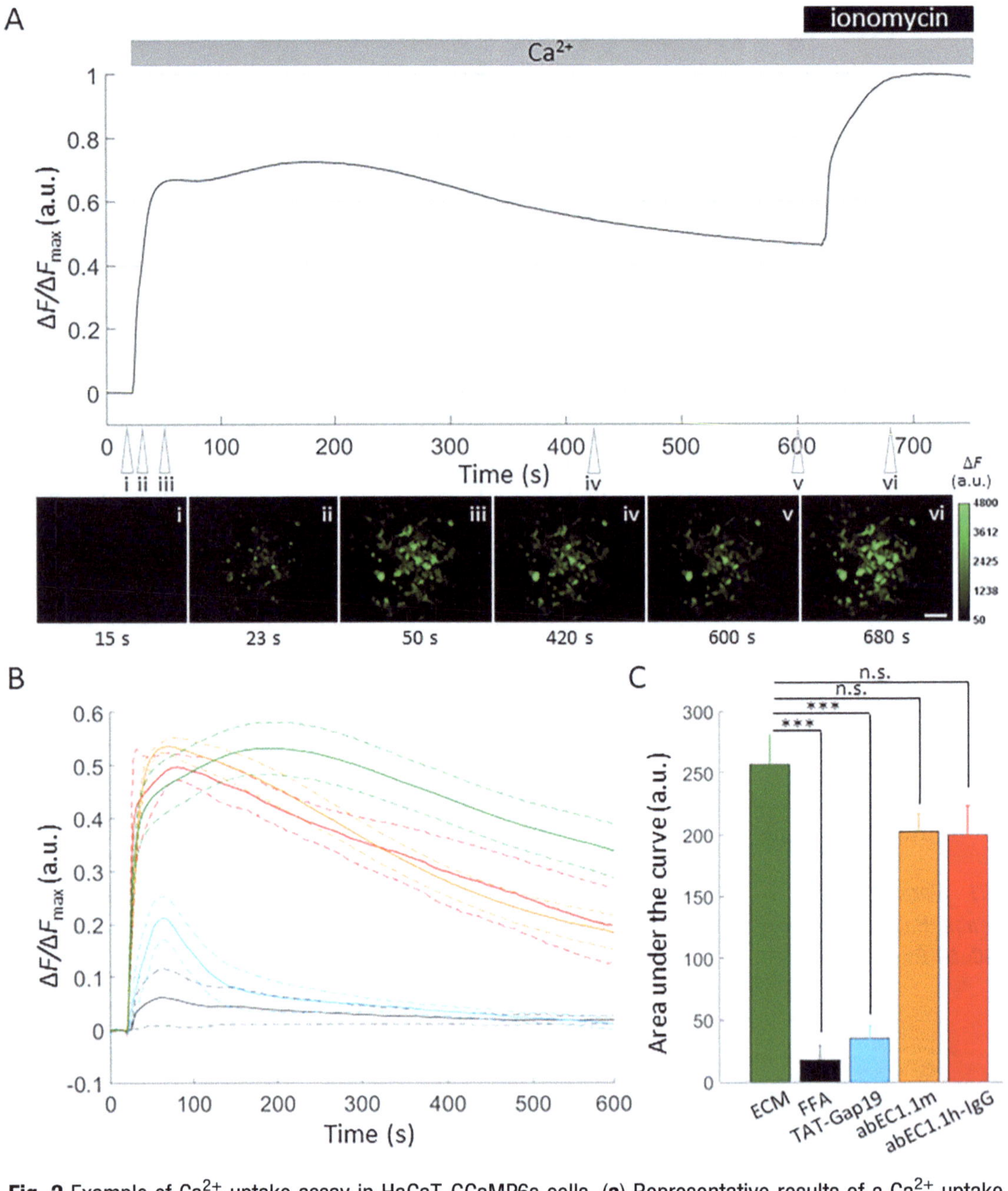

Fig. 2 Example of Ca^{2+} uptake assay in HaCaT-GCaMP6s cells. (**a**) Representative results of a Ca^{2+} uptake experiment on a single cell culture. Top: normalized Ca^{2+} uptake trace obtained by averaging GCaMP6s fluorescence intensity variations, ΔF, in all above-threshold pixels in the FOV; horizontal bars mark the addition of Ca^{2+} (gray, bolus volume: 2 μL, concentration: 1 M, delivery time: $t = 20$ s) and ionomycin (black, bolus volume: 2 μL, concentration: 200 μM, delivery time: $t > 600$ s) to the extracellular medium ($t = 0$ s indicates onset of recording); the trace was normalized to the maximum value reached after ionophore addition, ΔF_{max}. Bottom: representative ΔF frames acquired during imaging experiment at the time points indicated by gray triangles on the graph *x*-axis and denoted with roman numbers; scale: 100 μm. (**b**) Mean normalized Ca^{2+} uptake traces (solid lines) ± standard error of the mean (dashed lines) obtained from *n* cell cultures kept in the following extracellular media (ECM): ECM (green, $n = 9$), ECM plus FFA (FFA, 100 μM, black, $n = 3$), ECM plus TAT-Gap19 (TAT-Gap19, 150 μM, light blue, $n = 3$), ECM plus abEC1.1m (abEC1.1m,

4. Cell culture medium: Dulbecco's modified Eagle's medium (DMEM), supplemented with 10% (*v/v*) fetal bovine serum (FBS), 1 mM sodium pyruvate, 4 mM L-glutamine, 25 mM D-glucose, 100 μg/mL streptomycin, and 100 units/mL penicillin. Store at +4 °C.
5. Opti-Minimal Essential Medium (Opti-MEM). Store at +4 °C.
6. Dulbecco's phosphate-buffered saline (DPBS): 8000 mg/L NaCl, 200 mg/L KCl, 2160 mg/L Na_2HPO_4-$7H_2O$, 200 mg/L KH_2PO_4 in double-distilled water, pH 7.0–7.3. Store at +4 °C.
7. Trypsin-EDTA (0.25%). Store at +4 °C when performing experiments on a regular basis, otherwise store at −20 °C.
8. Poly-L-lysine stock solution: dissolve poly-L-lysine hydrobromide at a concentration of 1 mg/mL in double-distilled water. Aliquot and store at +4 °C when performing experiments on a regular basis, otherwise store at −20 °C.
9. Lipofectamine 2000. Store at +4 °C.
10. 10 cm diameter tissue culture dishes.
11. Polyethersulfone (PES) membrane filters (pore size = 0.45 μm).
12. Cryovials.
13. Ice in an ice bucket.
14. Humidified incubator (37 °C, 90% humidity, 5% CO_2).
15. Class 2 biosafety cabinet.
16. Ultracentrifuge.
17. Inverted fluorescence microscope equipped with green fluorescence filters.

2.1.2 HaCaT Cell Transduction

1. HaCaT cells.
2. LV encoding for GCaMP6s biosensor.
3. DPBS.
4. Trypsin-EDTA (0.25%).
5. Cell culture medium (*see* Subheading 2.1.1, **Item 4**).

Fig. 2 (continued) 1 μM, orange, $n = 3$) and ECM plus abEC1.1h-IgG (abEC1.1h-IgG, 1 μM, red, $n = 3$). (**c**) Quantification of cytosolic Ca^{2+} load for results shown in (**b**) as the area subtended by normalized Ca^{2+} uptake trace in the time interval between $t = 20$ s (Ca^{2+} bolus delivery) and $t = 590$ s (before ionomycin addition). Data are quoted as mean ± standard error of the mean. Statistical significance was calculated using Bonferroni *post hoc* test for pairwise comparisons after ANOVA test. *n.s.* not significant, ***p-value < 0.001

6. Polybrene stock solution: dissolve polybrene at a concentration of 4 mg/mL in double-distilled water. Aliquot and store at +4 °C when performing experiments on a regular basis, otherwise store at −20 °C.
7. 35 mm diameter tissue culture dish.
8. Ice in an ice bucket.
9. Humidified incubator (37 °C, 90% humidity, 5% CO_2).
10. Class 2 biosafety cabinet.
11. Inverted fluorescence microscope equipped with green fluorescence filters.

2.2 Measurement of Ca^{2+} Uptake

1. HaCaT-GCaMP6s cells.
2. DPBS.
3. Trypsin-EDTA (0.25%).
4. Cell culture medium (*see* Subheading 2.1.1, **Item 4**).
5. Sterilized round glass coverslips (12 mm diameter).
6. Sterilized forceps.
7. Coating medium: dilute laminin to a concentration of 10 μg/mL in sterile double-distilled water (*see* **Note 2**).
8. Ice in an ice bucket.
9. 35 mm diameter dish.
10. Extracellular medium (ECM): 138 mM NaCl, 5 mM KCl, 0.06 mM $CaCl_2$, 0.4 mM NaH_2PO_4, 6 mM D-glucose, and 10 mM HEPES in double-distilled water. Adjust pH with HCl/NaOH to 7.3. Filter-sterilize and store at +4 °C when performing experiments on a regular basis, otherwise store at −20 °C.
11. $CaCl_2$ bolus medium: 1 M $CaCl_2$ in double-distilled water. Store at +4 °C.
12. Ionomycin stock solution: dissolve ionomycin calcium salt at a concentration of 1 mM in dimethyl sulfoxide (DMSO). Aliquot and store at −20 °C for up to 1 year (*see* **Note 3**).
13. Ionomycin bolus medium: dilute the ionomycin stock solution at a concentration of 200 μM in ECM.
14. FFA stock solution: dissolve FFA at a concentration of 100 mM in DMSO. Aliquot and store at −20 °C for a maximum of 10 days.
15. FFA-containing solution: dilute the FFA stock solution in ECM to a final concentration of 100 μM.
16. TAT-Gap19 stock solution: dissolve TAT-Gap19 (amino acid sequence, from N- to C-terminus, YGRKKRRQRRRK QIEIKKFK) at a concentration of 1 mg/mL in DPBS. Aliquot and store at −20 °C for up to 2 years.

17. TAT-Gap19-containing solution: dilute the TAT-Gap19 stock solution in ECM to a final concentration of 150 μM.
18. abEC1.1 stock solutions: dilute abEC1.1m or abEC1.1h-IgG at a concentration of 20 μM in DPBS. Aliquot and store abEC1.1m at +4 °C and abEC1.1h-IgG at −20 °C for up to 1 year.
19. abEC1.1-containing solutions: dilute the abEC1.1 stock solutions in ECM to a final concentration of 1 μM.
20. 0.2–2 μL manual single-channel pipette.
21. Upright (spinning-disk) fluorescence microscope for live-cell imaging equipped with 20× water immersion objective, live-cell imaging chamber with 1 mL capacity, cooled sCMOS camera, 488 nm diode laser and 535/30 nm fluorescence emission band-pass filter to respectively excite and collect GCaMP6s fluorescence signal.
22. Humidified incubator (37 °C, 90% humidity, 5% CO_2).
23. Class 2 biosafety cabinet.

3 Methods

3.1 Generation of Stable HaCaT-GCaMP6s Cell Line

3.1.1 LV Production

1. Maintain the HEK 293T cells in cell culture medium at 37 °C with a constant supply of 5% CO_2.
2. Produce LV from pHAGE-RSV-GCaMP6s plasmid according to standard protocols (*see* Subheading 1) using HEK 293T cells.
3. In case of lipofectamine-mediated transfection in a 10 cm-diameter tissue culture dish, prepare the transfection mixture as follows: 60 μL of Lipofectamine 2000, 15 μg of lentiviral vector plasmid pHAGE-RSV-GCaMP6s, and 7.5 μg of each packaging vector, pMD2.G and psPAX2, in a total volume of 1.5 mL Opti-MEM.
4. For the following steps, either use LV right after production, or store aliquots at −80 °C for later usage.

3.1.2 HaCaT Cell Transduction

1. Maintain parental HaCaT cells in cell culture medium at 37 °C with a constant supply of 5% CO_2.
2. The day before lentiviral transduction, plate 2.5×10^5 cells in a 35 mm diameter tissue culture dish and incubate in the humidified incubator overnight.
3. On the day of lentiviral transduction, wash plated HaCaT cells once with DMEM to remove traces of FBS.

4. Dilute the LV (*see* **Note 4**) encoding for GCaMP6s in a total volume of 0.5 mL of DMEM (no FBS, no antibiotics) at a concentration of 2.5×10^6 transducing units/mL (*see* **Note 5**).
5. Add 1 μL of polybrene stock solution to 0.5 mL of LV-containing DMEM (final polybrene concentration: 8 μg/mL).
6. Remove the DMEM from the dish containing plated HaCaT cells and replace it with the LV-containing medium.
7. Incubate plated HaCaT cells with the LV-containing medium in the humidified incubator for 5 h (*see* **Note 6**).
8. Add 1.5 mL of cell culture medium to the dish and incubate overnight.
9. The next day, replace LV-containing medium with 2 mL of fresh cell culture medium.
10. Two days after lentiviral transduction, check for the presence of green fluorescence signal in transduced cells (*see* **Note 7**).
11. Culture HaCaT-GCaMP6s cells in normal cell culture medium at 37 °C with a constant supply of 5% CO_2.

3.2 Measurement of Ca^{2+} Uptake

1. The day before the experiment, prepare fresh coating medium (*see* **Note 2**).
2. Using sterilized forceps, place 3–5 round-glass coverslips into a 35 mm diameter dish and submerge in 2 mL of fresh coating medium (*see* **Note 8**).
3. Place the coverslip-containing dish in the humidified incubator for 30 min, then remove the coating medium.
4. Plate HaCaT-GCaMP6s cells on the laminin-coated coverslips at 80% confluence and incubate overnight in the humidified incubator.
5. On the day of the experiment, prepare fresh ionomycin bolus medium and inhibitor-containing solutions (FFA, TAT-Gap19, abEC1.1m or abEC1.1h-IgG) (*see* **Note 9**).
6. Before performing the Ca^{2+} uptake measurement, set imaging parameters (laser power, camera exposure time) to avoid GCaMP6s signal saturation during the experiments (*see* **Note 10**).
7. To measure Ca^{2+} uptake through HCs, transfer a cell-plated coverslip to the live-cell imaging chamber on the stage of the (spinning disk) fluorescence microscope and incubate it in 1 mL of ECM for 10 min at room temperature.
8. Alternatively, to evaluate the efficacy of HC inhibitors, incubate the cell-plated coverslip with the desired inhibitor-containing solution in the humidified incubator (40 min for FFA, 20 min for TAT-Gap19, 30 min for abEC1.1m or abEC1.1h-IgG),

then transfer the coverslip to the live-cell imaging chamber on the stage of the (spinning disk) fluorescence microscope and incubate it in 1 mL of inhibitor-containing solution for 5 min at room temperature.

9. Focus on HaCaT-GCaMP6s cells in the FOV and start a protocol for the acquisition of fluorescence images at 1 frame/s for a total time of 12 min.
10. After 20 s from the start of image acquisition, rapidly deliver 2 μL of $CaCl_2$ bolus medium at the edge of the FOV using a manual pipette (0.2–2 μL volume) while continuing image acquisition at unchanged frame rate (*see* **Note 11**).
11. After 10 min from the onset of the imaging protocol, rapidly deliver 2 μL of ionomycin bolus medium (same as above) (*see* **Note 12**).
12. At the end of the imaging protocol, discard the sample.
13. Before starting a new experiment on another sample, abundantly wash the live-cell imaging chamber and the objective lens with double-distilled water to remove residual ionomycin solution (*see* **Note 13**).
14. For each experiment, extract the normalized Ca^{2+} uptake trace from the acquired image sequence: generate a single $\Delta F = F - F_0$ trace (F_0 = GCaMP6s fluorescence emission averaged over all prestimulus frames) by spatially averaging pixel signals over all HaCaT-GCaMP6s cells in the FOV; normalize the ΔF trace for the peak response to ionomycin, ΔF_{max}, and use the $\Delta F/\Delta F_{max}$ trace thus obtained for statistical analyses.

4 Notes

1. Using cells of low passage number is critical for efficient LV production.
2. Carefully thaw laminin on ice. Avoid multiple freeze-thaw cycles. Keep the coating medium on ice until used.
3. Use DMSO under a properly functioning chemical fume hood.
4. In case LV was previously prepared and stored at −80 °C, slowly thaw a LV aliquot on ice. Avoid multiple freeze-thaw cycles to preserve LV titer.
5. For HaCaT cell transduction, a multiplicity of infection (MOI) around 10 is recommended.
6. In the first 30 min of incubation, periodic gentle shaking of the dish every 5–10 min is recommended.
7. The use of an inverted fluorescence microscope allows to maintain cell culture sterility. Replacing cell culture medium with DPBS may facilitate inspection at the microscope.

8. The laminin coating concentration is about 2 μg/cm^2.
9. Unused ionomycin bolus medium and abEC1.1h-IgG-containing solution can be frozen at −20 °C, thawed on ice and used more than once (use frozen ionomycin bolus medium within a month). abEC1.1m-containing solution can be stored at +4 °C and used more than once. Unused FFA- and TAT--Gap19-containing solutions must be discarded.
10. This is a trial-and-error procedure in which every sample can be used only for one trial. Transfer a coverslip with adherent HaCaT-GCaMP6s cells to the live-cell imaging chamber on the stage of the (spinning disk) fluorescence microscope and incubate the cells in ECM at room temperature. Check that, with chosen imaging parameters, HaCaT-GCaMP6s Ca^{2+} responses to ionomycin bolus medium are not saturated.
11. The final estimated $[Ca^{2+}]_{ex}$ is 2 mM. Bolus delivery should be precise and rapid, possibly performed between two consecutive exposures of the microscope camera. In the radial direction with respect to the object optical axis, the pipette tip must be placed near the edge of the objective lens. In the axial direction, the pipette tip must be neither too close to the sample, to avoid cell movement or damage, nor in contact with the lens, to avoid damage to the objective and motion artifacts in the recordings due to ECM flux induced by bolus delivery. We recommend forming an angle between the pipette tip direction and the objective optical axis of about 40°. An expert experimenter can deliver the 2 μL bolus without introducing motion artifacts. In case a frame affected by ECM motion artifact is present, it can be deleted offline from the image series before starting image postprocessing and analysis.
12. The final estimated ionomycin concentration is 0.4 μM. Ionomycin is a Ca^{2+} ionophore that equilibrates the intracellular and extracellular free Ca^{2+}concentrations, here used to normalize the Ca^{2+} influx response to the maximum achievable fluorescence variation in that particular cell population, that is, to the level of biosensor expression throughout experiments or experimental sessions. Peak response to ionomycin is typically reached within 50–90 s after bolus delivery. In this time window, no visible signs of cell stress are observed. Occasionally, some plasma membrane blebbing could occur towards the end of the imaging protocol, but it does not affect the determination of the maximum achievable GCaMP6s fluorescence intensity.
13. This is a critical step: residual traces of ionomycin used in the previous experiment may alter the intracellular Ca^{2+} concentration of cells following $CaCl_2$ bolus medium delivery in the next experiment and determine an overestimation of Ca^{2+}

uptake. If some membrane blebbing is noticed (typically in the marginal portions of the FOV) before the delivery of the ionomycin bolus medium, stop the experiment, discard the sample and repeat the washing. Remove the objective from the microscope to wash the lens properly. After washing with running water, the live-cell imaging chamber can be dipped in double-distilled water for 5–10 min, before drying and reuse.

Acknowledgments

This work was funded by grants awarded to Fabio Mammano by Fondazione Telethon (GGP19148), the Italian Ministry of Education, Universities and Research (PRIN 20175C22WM) and the University of Padova (Prot. BIRD187130).

References

1. Maeda S, Nakagawa S, Suga M et al (2009) Structure of the connexin 26 gap junction channel at 3.5 A resolution. Nature 458(7238):597–602. https://doi.org/10.1038/nature07869
2. Myers JB, Haddad BG, O'Neill SE et al (2018) Structure of native lens connexin 46/50 intercellular channels by cryo-EM. Nature 564(7736):372–377. https://doi.org/10.1038/s41586-018-0786-7
3. Lee HJ, Jeong H, Hyun J et al (2020) Cryo-EM structure of human Cx31.3/GJC3 connexin hemichannel. Sci Adv 6(35):eaba4996. https://doi.org/10.1126/sciadv.aba4996
4. Khan AK, Jagielnicki M, Bennett BC et al (2021) Cryo-EM structure of an open conformation of a gap junction hemichannel in lipid bilayer nanodiscs. Structure. https://doi.org/10.1016/j.str.2021.05.010
5. Wang N, De Bock M, Decrock E et al (2013) Paracrine signaling through plasma membrane hemichannels. Biochim Biophys Acta 1828(1):35–50. https://doi.org/10.1016/j.bbamem.2012.07.002
6. Retamal MA, Reyes EP, Garcia IE et al (2015) Diseases associated with leaky hemichannels. Front Cell Neurosci 9:267. https://doi.org/10.3389/fncel.2015.00267
7. Laird DW, Lampe PD (2022) Cellular mechanisms of connexin-based inherited diseases. Trends Cell Biol 32(1):58–69. https://doi.org/10.1016/j.tcb.2021.07.007
8. Schalper KA, Palacios-Prado N, Retamal MA et al (2008) Connexin hemichannel composition determines the FGF-1-induced membrane permeability and free [Ca2+]i responses. Mol Biol Cell 19(8):3501–3513. https://doi.org/10.1091/mbc.E07-12-1240
9. Schalper KA, Sanchez HA, Lee SC et al (2010) Connexin 43 hemichannels mediate the Ca2+ influx induced by extracellular alkalinization. Am J Physiol Cell Physiol 299(6):C1504–C1515. https://doi.org/10.1152/ajpcell.00015.2010
10. Fiori MC, Figueroa V, Zoghbi ME et al (2012) Permeation of calcium through purified connexin 26 hemichannels. J Biol Chem 287(48):40826–40834. https://doi.org/10.1074/jbc.M112.383281
11. Sanchez HA, Orellana JA, Verselis VK et al (2009) Metabolic inhibition increases activity of connexin-32 hemichannels permeable to Ca2+ in transfected HeLa cells. Am J Physiol Cell Physiol 297(3):C665–C678
12. Mandal A, Shahidullah M, Delamere NA (2015) Calcium entry via connexin hemichannels in lens epithelium. Exp Eye Res 132:52–58. https://doi.org/10.1016/j.exer.2015.01.012
13. Paul DL, Ebihara L, Takemoto LJ et al (1991) Connexin46, a novel lens gap junction protein, induces voltage-gated currents in nonjunctional plasma membrane of Xenopus oocytes. J Cell Biol 115(4):1077–1089
14. Ebihara L, Steiner E (1993) Properties of a nonjunctional current expressed from a rat connexin46 cDNA in Xenopus oocytes. J Gen Physiol 102(1):59–74
15. Contreras JE, Saez JC, Bukauskas FF et al (2003) Gating and regulation of connexin

43 (Cx43) hemichannels. Proc Natl Acad Sci USA 100(20):11388–11393. https://doi.org/10.1073/pnas.1434298100

16. Bennett MV, Contreras JE, Bukauskas FF et al (2003) New roles for astrocytes: gap junction hemichannels have something to communicate. Trends Neurosci 26(11):610–617
17. Goodenough DA, Paul DL (2003) Beyond the gap: functions of unpaired connexon channels. Nat Rev Mol Cell Biol 4(4):285–294
18. Saez JC, Retamal MA, Basilio D et al (2005) Connexin-based gap junction hemichannels: gating mechanisms. Biochim Biophys Acta 1711(2):215–224. https://doi.org/10.1016/j.bbamem.2005.01.014
19. Fasciani I, Temperan A, Perez-Atencio LF et al (2013) Regulation of connexin hemichannel activity by membrane potential and the extracellular calcium in health and disease. Neuropharmacology 75:479–490. https://doi.org/10.1016/j.neuropharm.2013.03.040
20. Saez JC, Leybaert L (2014) Hunting for connexin hemichannels. FEBS Lett 588(8):1205–1211. https://doi.org/10.1016/j.febslet.2014.03.004
21. Leybaert L, Lampe PD, Dhein S et al (2017) Connexins in cardiovascular and neurovascular health and disease: pharmacological implications. Pharmacol Rev 69(4):396–478. https://doi.org/10.1124/pr.115.012062
22. Garcia IE, Bosen F, Mujica P et al (2016) From hyperactive Connexin26 hemichannels to impairments in epidermal calcium gradient and permeability barrier in the keratitis-ichthyosis-deafness syndrome. J Invest Dermatol 136(3):574–583. https://doi.org/10.1016/j.jid.2015.11.017
23. Retamal MA, Fernandez-Olivares A, Stehberg J (2021) Over-activated hemichannels: a possible therapeutic target for human diseases. Biochim Biophys Acta Mol Basis Dis 1867(11):166232. https://doi.org/10.1016/j.bbadis.2021.166232
24. Van Campenhout R, Gomes AR, De Groof TWM et al (2021) Mechanisms underlying connexin hemichannel activation in disease. Int J Mol Sci 22(7). https://doi.org/10.3390/ijms22073503
25. Willebrords J, Maes M, Crespo Yanguas S et al (2017) Inhibitors of connexin and pannexin channels as potential therapeutics. Pharmacol Ther 180:144–160. https://doi.org/10.1016/j.pharmthera.2017.07.001
26. Delvaeye T, Vandenabeele P, Bultynck G et al (2018) Therapeutic targeting of connexin channels: new views and challenges. Trends Mol Med 24(12):1036–1053. https://doi.org/10.1016/j.molmed.2018.10.005
27. Laird DW, Lampe PD (2018) Therapeutic strategies targeting connexins. Nat Rev Drug Discov 17(12):905–921. https://doi.org/10.1038/nrd.2018.138
28. Zhang C, Yan Z, Maknojia A et al (2021) Inhibition of astrocyte hemichannel improves recovery from spinal cord injury. JCI Insight 6(5). https://doi.org/10.1172/jci.insight.134611
29. Xu L, Carrer A, Zonta F et al (2017) Design and characterization of a human monoclonal antibody that modulates mutant connexin 26 hemichannels implicated in deafness and skin disorders. Front Mol Neurosci 10:298. https://doi.org/10.3389/fnmol.2017.00298
30. Ziraldo G, Buratto D, Kuang Y et al (2019) A human-derived monoclonal antibody targeting extracellular connexin domain selectively modulates hemichannel function. Front Physiol 10:392. https://doi.org/10.3389/fphys.2019.00392
31. Kuang Y, Zorzi V, Buratto D et al (2020) A potent antagonist antibody targeting connexin hemichannels alleviates Clouston syndrome symptoms in mutant mice. EBioMedicine 57:102825. https://doi.org/10.1016/j.ebiom.2020.102825
32. Buratto D, Donati V, Zonta F et al (2021) Harnessing the therapeutic potential of antibodies targeting connexin hemichannels. Biochim Biophys Acta Mol Basis Dis 1867(4):166047. https://doi.org/10.1016/j.bbadis.2020.166047
33. Boukamp P, Petrussevska RT, Breitkreutz D et al (1988) Normal keratinization in a spontaneously immortalized aneuploid human keratinocyte cell line. J Cell Biol 106(3):761–771
34. Fitzgerald DJ, Fusenig NE, Boukamp P et al (1994) Expression and function of connexin in normal and transformed human keratinocytes in culture. Carcinogenesis 15(9):1859–1865. https://doi.org/10.1093/carcin/15.9.1859
35. Nugent L, Ofori-Frimpong B, Martin PE et al (2021) Cx31.1 expression is modulated in HaCaT cells exposed to UV-induced damage and scrape-wounding. J Cell Physiol 236(2):911–920. https://doi.org/10.1002/jcp.29901
36. Qin Y, Han L, Yang D et al (2018) Silver nanoparticles increase connexin43-mediated gap junctional intercellular communication in HaCaT cells through activation of reactive oxygen species and mitogen-activated protein kinase signal pathway. J Appl Toxicol 38(4):564–574. https://doi.org/10.1002/jat.3563

37. Zhang Q, Wu S, Liu L et al (2019) Effects of bisphenol A on gap junctions in HaCaT cells as mediated by the estrogen receptor pathway. J Appl Toxicol 39(2):271–281. https://doi.org/10.1002/jat.3717
38. Garcia-Vega L, O'Shaughnessy EM, Jan A et al (2019) Connexin 26 and 43 play a role in regulating proinflammatory events in the epidermis. J Cell Physiol. https://doi.org/10.1002/jcp.28206
39. Johnson RG, Le HC, Evenson K et al (2016) Connexin Hemichannels: methods for dye uptake and leakage. J Membr Biol 249(6):713–741. https://doi.org/10.1007/s00232-016-9925-y
40. Anselmi F, Hernandez VH, Crispino G et al (2008) ATP release through connexin hemichannels and gap junction transfer of second messengers propagate Ca2+ signals across the inner ear. Proc Natl Acad Sci USA 105(48):18770–18775. https://doi.org/10.1073/pnas.0800793105
41. Nardin C, Peres C, Putti S et al (2021) Connexin hemichannel activation by S-nitrosoglutathione synergizes strongly with photodynamic therapy potentiating antitumor bystander killing. Cancers 13(20). https://doi.org/10.3390/cancers13205062
42. Chiara N, Abraham T-M, Viola D et al (2022) A Quantitative Assay for Ca2+ Uptake through Normal and Pathological Hemichannels. Int J Mol Sci 23(13):7337. https://doi.org/10.3390/ijms23137337
43. Chen TW, Wardill TJ, Sun Y et al (2013) Ultrasensitive fluorescent proteins for imaging neuronal activity. Nature 499(7458):295–300. https://doi.org/10.1038/nature12354
44. Cribbs AP, Kennedy A, Gregory B et al (2013) Simplified production and concentration of lentiviral vectors to achieve high transduction in primary human T cells. BMC Biotechnol 13:98. https://doi.org/10.1186/1472-6750-13-98
45. Ceriani F, Ciubotaru CD, Bortolozzi M et al (2016) Design and construction of a cost-effective spinning disk system for live imaging of inner ear tissue. Methods Mol Biol 1427:223–241. https://doi.org/10.1007/978-1-4939-3615-1_13
46. Abudara V, Bechberger J, Freitas-Andrade M et al (2014) The connexin43 mimetic peptide Gap19 inhibits hemichannels without altering gap junctional communication in astrocytes. Front Cell Neurosci 8:306. https://doi.org/10.3389/fncel.2014.00306

Chapter 9

Evaluation of Connexin Hemichannel Activity In Vivo

Francisca M. Acosta, Manuel A. Riquelme, Sumin Gu, and Jean X. Jiang

Abstract

Connexin hemichannels (Cx HCs) are hexameric structures at the cell plasma membrane, whose function as membrane transport proteins allows for the passive flow of small hydrophilic molecules and ions ($\leq$1 kDa) between the cytosol and the extracellular environment. Activation of Cx HCs is highly dependent on pathological conditions. HC activity provokes changes in the microenvironment, inducing the dissemination of signaling molecules in both an autocrine and paracrine manner. Given the elicitation of a variety of signaling pathways, and assortment of Cx species and dispersion throughout the body, Cx HCs have been implicated in a range of processes such as cell proliferation, differentiation, cell death, and tissue modeling and remodeling. While studying the expression and localization of Cx HCs can be done using traditional laboratory techniques, such as immunoblot analysis, measuring the functionality/activity of the HCs requires a more explicit methodology and is essential for determining Cx-mediated physiological changes. The study of Cx HC function/activity has focused mainly on in vitro measurements through electrophysiological characterization or, more commonly, using HC-permeable dye uptake studies. Here, we describe the use of dye uptake to measure Cx HC activity in vivo using mechanically stimulated osteocytic Cx43 HCs with Evans blue dye as our model.

Key words Connexin hemichannel, Dye uptake, Pathology, In vivo, Cx43

1 Introduction

Hemichannels (HCs) from two adjoining cells can dock head-to-head, forming a gap junction channel that mediates cytoplasmic communication, while undocked HCs can also serve as sovereign communication conduits for extracellular communication between the cytosol and the extracellular environment [1, 2]. A connexin (Cx) HC is formed by six Cx monomers. In humans, 21 Cx species have been identified distributed in almost all organs and cell types, with some exceptions being red blood cells and healthy mature skeletal muscle myofibers [2]. Members of the Cx family are named based upon their calculated molecular weight; for example, connexin 43 (Cx43), the most abundant connexin isoform, has a molecular weight of 43 kDa [2].

Fabio Mammano and Mauricio Retamal (eds.), *Connexin Hemichannels: Methods and Protocols*, Methods in Molecular Biology, vol. 2801, https://doi.org/10.1007/978-1-0716-3842-2_9,

Autocrine and paracrine cellular communication mediated by Cx HCs facilitates the diffusion of ions and metabolites. HCs are permeable to ions, such as sodium (Na^+), potassium (K^+), calcium (Ca^{2+}), and signaling molecules $\leq$1 kDa, such as adenosine triphosphate (ATP), prostaglandin E_2 (PGE_2), glutamate, and other small molecules such as dyes with low molecular weight [3, 4]. Under normal physiological conditions, undocked Cx HCs remain mainly closed; however, pathophysiological conditions, such as during injury or disease, drive their opening, promoting the extracellular release of HC permeable molecules, thus they are known to be major regulators of cellular homeostasis and disease [5]. Among the mechanisms underlying Cx HC activation are alkaline pH, hypoxia, metabolic inhibition, oxidative stress, temperature, and low extracellular Ca^{2+} [6–8]. Practices such as membrane depolarization, mechanical loading, the introduction of cytokines/growth factors, or deliberate increase of intracellular calcium concentration are also means of stimulating Cx HC opening [9–11].

Given this perspective, when studying Cx HCs in the context of pathophysiological development and cellular homeostasis, beyond the measurement of Cx HC expression and localization, evaluation of HC activity/function is critical. These analytical measurements are commonly done through the use of electrophysiological characterization or, more commonly, using HC-permeable dye uptake studies [3, 12]. In particular, dye uptake has been broadly used as a means of measuring in vitro plasma membrane permeability with charged dyes, such as ethidium bromide, DAPI, Lucifer yellow, Evans blue, among others. These dyes, in conjunction with the addition of specific blockers, could be used to measure the partial participation of Cx HC activity in the increase of membrane permeability [13–15]. The principle behind this technique is that these charged hydrophilic fluorescent dyes can be added in high concentrations extracellularly and cannot innately cross the plasma membrane. In the presence of stimulation, activating or opening HCs, like those mentioned above, the dye diffuses into the cytoplasm through functional and permeable Cx HCs. Excess extracellular dye is then rinsed off, leaving behind dye trapped in the interior of the cells. The incorporated dye can then be quantified using fluorescence microscopy [14]. Explicitly, images containing different fields of view are captured and quantified, utilizing the pixel fluorescence intensity of the dye captured, within the cells of interest, as an index of an increase in the plasma membrane permeability, or Cx HC activity [14].

This same approach, utilizing dye uptake as a means of measuring Cx HC activity/function, can be transferred to measure in vivo Cx HC activity [13, 16–18]. This powerful technique, when coupled with Cx specific inhibitors, mice models with genetic ablations, or different stimulatory techniques, can be used to study not only the activity/function of these HCs using a physiologically

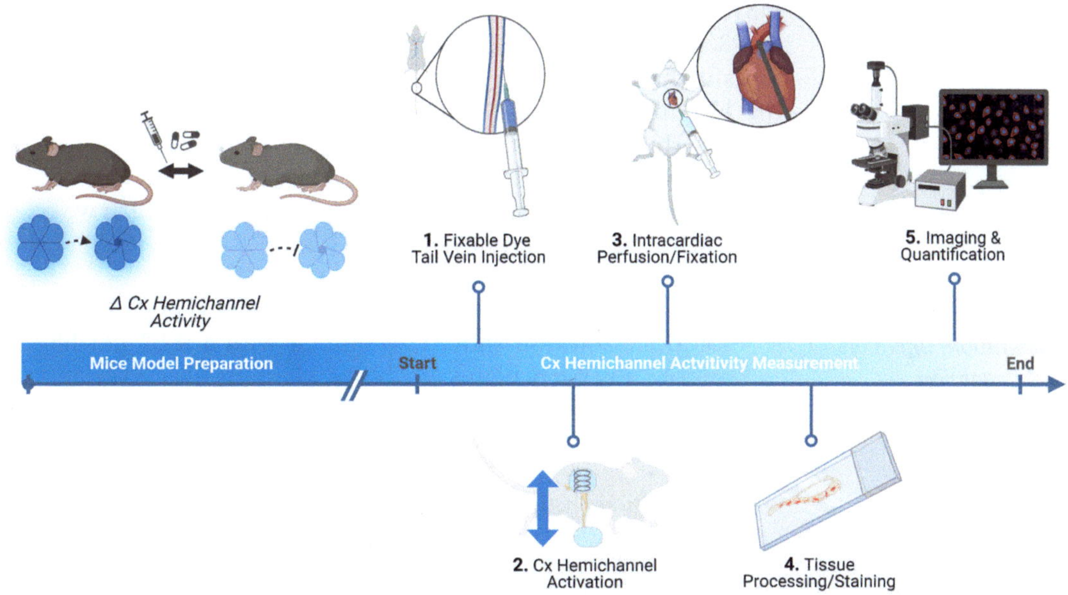

Fig. 1 Overview of process for in vivo quantification of Cx HC activity, starting at the point of postmodel development

relevant approach, but also couple this with local and global changes in pathophysiology. Here, we demonstrate this method using mechanically stimulated Cx43 HCs, in osteocytes, with Evans blue dye as our model. The advantage of using this model is that bone osteocytes primarily express Cx43 and unlike gap junction channels forming at narrow tips of dendrites, Cx43 HCs are richly present on the osteocyte cells surface [19, 20].

Cx43 HC activity was analyzed following (1) intravenous injection of Evans blue dye in mice; (2) activation of HCs by tibial cyclic axial compressions; (3) intracardiac perfusion and tissue fixation; (4) collection of tibial bones, processing, and cryosectioning; and (5) measurement of the fluorescence intensities of osteocytes, indicative of dye infiltration and HC activity, using NIH ImageJ software (Fig. 1).

2 Materials

Mice utilized for the study should ideally contain groups with potential differences in Cx HCs including the treatment with Cx HC specific inhibitors, mouse models containing genetic ablations, and/or the utilization of different stimulatory techniques, in order to obtain quantifiable Cx HC activity changes. For the example data presented here, Cx43 HCs in osteocytes were the target and mechanical stimulation, through tibia loading, was utilized in WT and transgenic mice (R76W and Δ130–136), driven by a 10-kb

DMP1 promoter with the overexpression of dominant negative Cx43 mutants in osteocytes. The R76W is a dominant negative mutant that blocks the endogenous gap junction channel, but not the HC function, and the dominant negative Δ130–136 mutant inhibits the activity of both types of channels [21]. Using these models, we have reported differences in bone anabolic function in response to mechanical loading/Cx43 HC activity, which we quantified utilizing this methodology [18].

2.1 Evans Blue Dye Preparation, Injection, and Collection

1. Dissolve 20 mg/mL Evans blue (EB) dye in sterile saline solution. Solution should be filtered with a 0.22 μm filter to remove any particulates (*see* **Note 1**).
2. 0.22 μm filter.
3. Scale.
4. Antiseptic solution (e.g., 70% ethanol (EtOH)).
5. Gauze pads.
6. 0.5 mL 25G insulin syringe.
7. Warmed phosphate-buffered saline (PBS) pH 7.4.
8. Mice mechanical restraint device.
9. Isoflurane.
10. Precision vaporizer system with, oxygen, anesthesia mask, and induction chamber.
11. 4% paraformaldehyde (PFA).
12. Surgical scissors, forceps, and hemostats.
13. Blunt 15G perfusion needle.
14. 3 × —100 mL syringe.
15. Peristaltic pump (*optional*).

2.2 Tibial Mid-diaphyseal Strain Measurements and Cyclic Tibial Loading

1. Strain gauge.
2. Bose ElectroForce LM1 test instrument.
3. Bridge completion module.
4. Signal conditioner/amplifier systems.
5. Tibial loading Device (Custom Apparatus).
6. Isoflurane.
7. Precision vaporizer system with oxygen, anesthesia mask, and induction chamber.

2.3 Cryosectioning, Imaging, and Quantification

1. 10% ethylenediaminetetraacetic acid (EDTA) in PBS.
2. 15% sucrose in PBS.
3. 30% sucrose in PBS.
4. Optimum cutting temperature compound (OCT).

5. Liquid nitrogen.
6. Cryostat.
7. Hydrophobic Barrier PAP Pen.
8. 4′,6-diami-dino-2-phenylindole (1 μg/mL, DAPI).
9. Fluoromount-G.
10. Optical fluorescence microscope.
11. NIH ImageJ software.

3 Methods

Carry out all procedures at room temperature unless otherwise specified. All animal protocols should be performed following the National Institutes of Health guidelines for care and use of laboratory animals and approved by Institutional Animal Care and Use Committees (IACUC).

3.1 Tibial Mid-diaphyseal Strain Measurements

1. (*see* **Note 2**) Complete this step prior to experiment set-up to ensure that proper calculation is made prior to study, to determine the applied load (N) required to generate equivalent micro strains (με) between your testing groups.
2. Euthanize mice, from each testing group, with isoflurane.
3. Exposed tibia bones using surgical instruments (*see* **Note 3**).
4. Attach strain gauge to the tibial diaphyseal anterior-medial surface (~37% down the tibia bone from the proximal end).
5. Using an electromagnetic testing system, such as the Bose ElectroForce LM1 with an attached bridge completion model and a conditioner/amplifier system, apply dynamic compressive loads from 0 to 9.5 N at the ends of the tibia.
6. Plot the slope between the load and micro strain ($N/\mu\varepsilon$), ($R^2 > 0.99$), indicative of the "Tibial Compliance" (Fig. 2).
7. Based on the calculated load–strain relationship, determine the force/load (N) to use for mechanical loading of groups (*see* **Note 4**).

3.2 Evans Blue Dye Tail Vein Injection

1. Preweigh mice using scale.
2. According to mouse weight, prepare syringes of EB solution (20 mg/mL in sterile saline) using insulin syringes, dye injected should be 200 mg/kg.
   (a) For example, for 20 g mouse, 200 μL of EB solution should be used.
3. Place mice in mechanical restraint of choice with tail vein protruding.

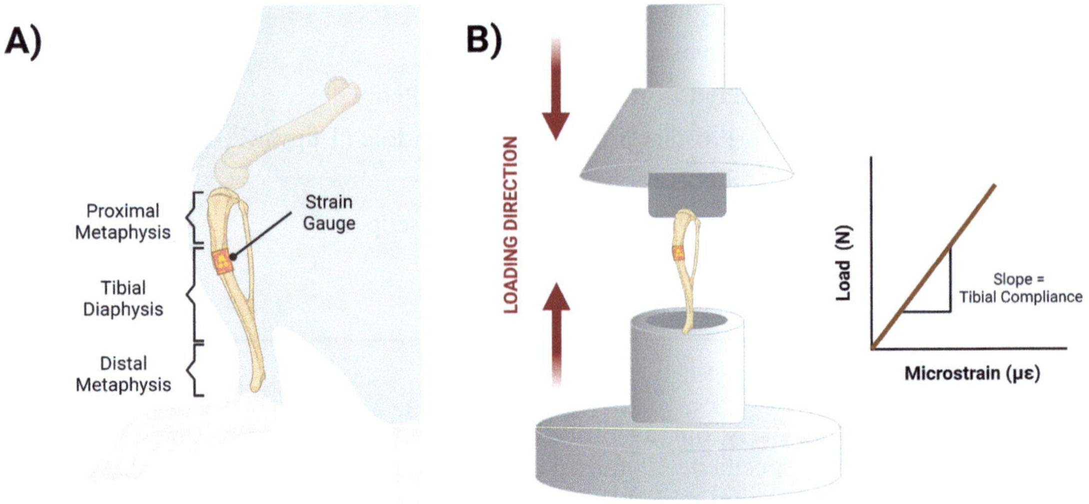

Fig. 2 Measurement of tibial mid-diaphyseal strain. (**a**) Placement of strain gauge in exposed tibial bone. (**b**) Setup for measurement of tibial compliance with an electromagnetic mechanical testing system

4. Wipe tail with antiseptic solution.
5. Immobilize tail vein with nondominant hand and rotate ~90° to access the lateral tail vein (*see* **Notes 5** and **6**).
6. Hold the tail taut with nondominant hand, near the tip of the tail, and insert the needle into the vein at a minimal angle (≤30°), to tail, with the beveled edge of needle facing upwards (*see* **Note** 7).
7. Advance the needle inward until ~ ½ the length is inside (~5 mm), careful as to not to perforate the vein (*see* **Note 8**).
8. Upon completion, remove needle and gently apply pressure at site of injection until bleeding has stopped (*see* **Note 9**).
9. Place animal back into housing container for 20 min then immediately proceed with tibial cyclic axial loading.
10. Repeat with any additional mice, as needed (Fig. 3).

3.3 Tibial Cyclic Axial Loading

1. Ensure proper weights are attached to the loading device based on calculated load–strain relationship (*see* **Note 10**), in addition to attaching a 0.4 N continuous static preload to hold tibia in place.
2. Preanesthetize mice by using a precision vaporizer system connected to an oxygen tank and placing mice inside the induction chamber and seal/close top for ~2–5 min at 2.5% isoflurane in 100% oxygen.
3. Ensure that animal has reached a surgical plane of anesthesia via toe pinch.

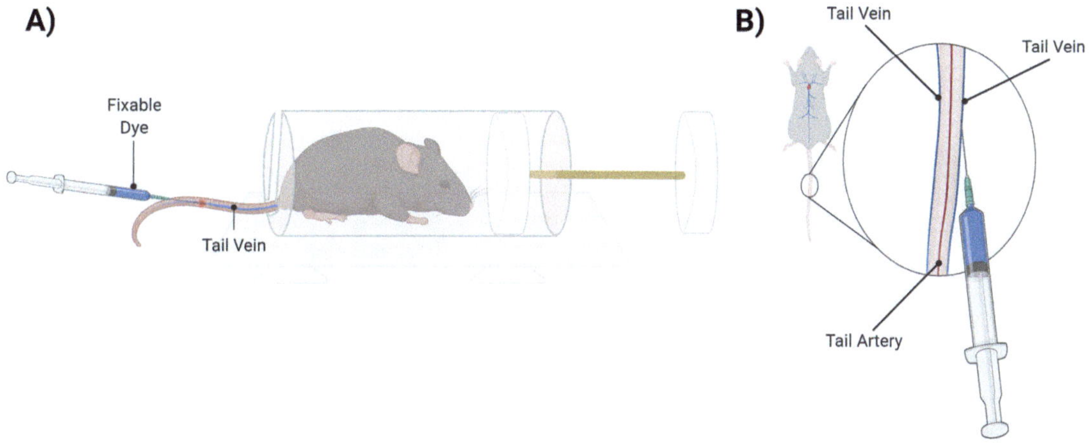

Fig. 3 Injection of mice tail vein with Evans blue dye for measurement of permeable Cx HCs. (**a**) Setup with mice in mechanical restraint with tail protruding and accessible for injection. (**b**) Anatomical landmarks needed for localization of tail vein

4. Switch system flow and transfer mice to an anesthesia mask/ nose cone in proximity to loading apparatus and adjust gas flow with vaporizer to 2–3% isoflurane.
5. Position the tibia to be loaded (*see* **Note 3**) into loading apparatus with an upper padded cup containing the knee connected to loading device, conducting the compressions, and a lower cup holding the heel in place.
6. Conduct the cyclic axial compressive loading (1200 cycles (10 min) at 2-Hz frequency with a sinusoidal waveform).
7. Place animal back into housing container for 40 min then immediately proceed with euthanization and perfusions.
8. Repeat with any additional mice, as needed (Fig. 4).

3.4 Euthanization and Perfusion

1. Euthanize mice with isoflurane.
2. Place the mouse on its back and firmly tape down its four paws to immobilize.
3. Using the surgical tools to perform a thoracotomy, cutting above the diaphragm and along the ribs up to the collarbone, to expose the heart, taking special care not to puncture the heart.
4. Carefully cut away any other tissues that may be occluding the heart to provide a clear view of the heart and vessels.
5. Connect a 15G blunt perfusion needle to a 100 mL syringe filled with PBS and flush the tubing and needle to expel air (*see* **Note 11**).
6. Pass the blunt 15G perfusion needle filled with PBS, through the tip of the left ventricle (*see* **Note 12**) at an angle parallel to

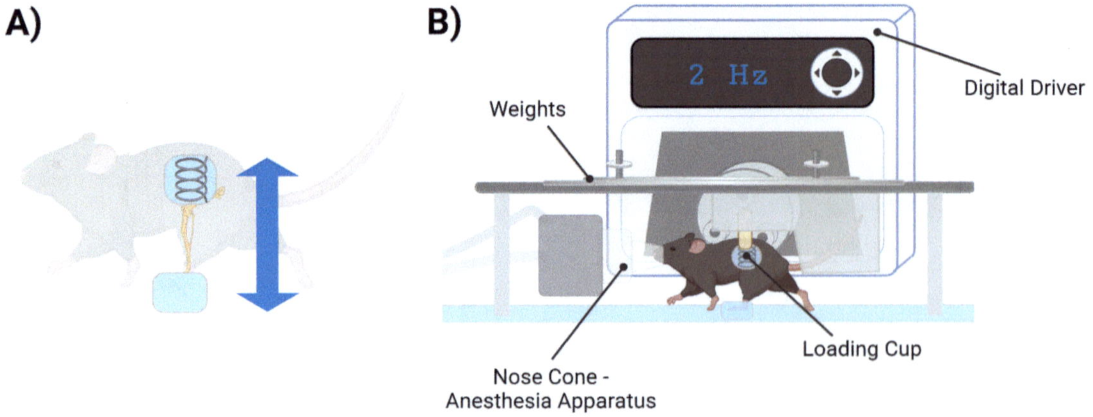

Fig. 4 Mice in vivo axial loading for activation of osteocyte Cx43 HCs. (**a**) Diagram of tibia positioning in the loading device and blue arrow demonstrating the direction of loading. (**b**) Schematic of the custom loading setup for tibial cyclic axial loading, specifically showing anesthetized mice undergoing loading in the left tibia using a predetermined force, to equalize strain

the midline of the heart until it enters the ascending aorta, visibly seen through aorta wall, but not reaching the aortic arch.

7. Using a hemostat, clamp the heart, to secure the needle and prevent leakage.
8. Make an incision on the right atrium using a small scissor to allow for the flow out of blood/liquid being perfused.
9. Begin PBS perfusion, to remove extraneous EB, by injecting warmed PBS at a constant speed of 4–5 mL per min for 10 min by manually pushing the syringe or using a peristaltic pump.
10. Switch from PBS to 4% PFA, making sure no bubbles get into the perfusion system, and perfuse mice, as before, to fix tissues.
11. For the last perfusion, switch from 4% PFA back to PBS, to completely administer the remaining PFA in the system, and perfuse as before (*see* **Note 13**).
12. Immediately proceed with isolation and processing of tibias.
13. Repeat with any additional mice, as needed (Fig. 5).

3.5 Isolation, Processing of Tibias, and Cryosectioning

1. Using the surgical instruments, isolate tibias taking special care in removing surrounding muscle and connective tissue.
2. Fix the tibias with 4% PFA for 2 days in 4 °C (in 10× tissue volume).
3. Rinse tibias with PBS three times for 30 min each.
4. Decalcify tibias with 10% ethylenediaminetetraacetic acid (EDTA) for 3 weeks (in 10× tissue volume) at 4 °C, changing out EDTA two times a week.

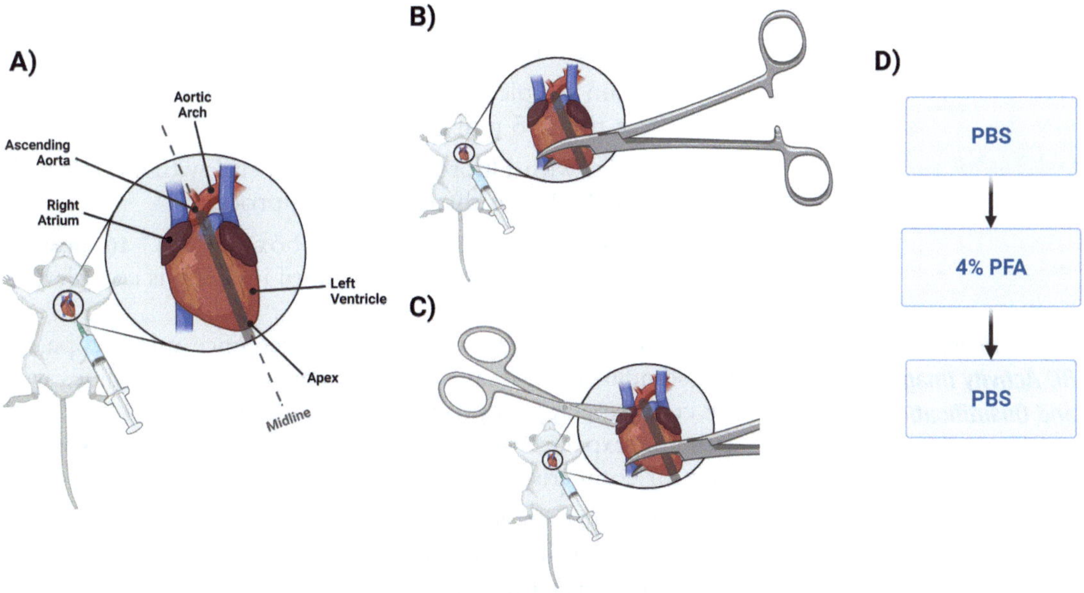

Fig. 5 Transcardiac perfusion of mice for tibial bone fixation. (**a**) Anatomical landmarks needed to perform procedure and position and direction of needle insertion. (**b**) Demonstration of the clamping of the needle to prevent movement and leakage. (**c**) Small incision being made in the right atrium to allow for the outflow of perfused liquids. (**d**) Flowchart of perfusion steps

5. Following decalcification, rinse tibias with PBS three times for 5 min each.
6. For cryopreservation, immerse decalcified tibias in 15% sucrose for 12 h at room temperature then 30% sucrose overnight at 4 °C (in 10× tissue volume).
7. Embed tibias longitudinally in OCT compound.
   (a) Dip the embedding mold repeatedly in liquid nitrogen for a short time until the OCT changes to a porcelain white color, indicating freezing of sample. Store at −20 °C for at least 12 h before cutting sections. Long-term storage can be done at −80 °C for ≤1 month (*see* **Note 14**).
8. Using a cryostat, set to −20 °C, cut longitudinal frozen tissue sections of 12 μm thickness, and place in a − 20 °C freezer until ready for staining.

3.6 Tibia Staining and Imaging

1. Remove slides from freezer and let samples come to room temperature for ~30 min.
2. Immerse the frozen tissue sections in cold acetone, which increases the permeability and fixation, at −20 °C for 2 min, and then air-dry the sections.
3. Use Hydrophobic Barrier PAP Pen to draw a liquid barrier around each tissue section.

4. After drying, rinse the slide with PBS, to remove residual OCT, washing three times, 5 min for each time.
5. Counterstain sample with DAPI for 10 min, followed by a PBS wash three times, 5 min each time.
6. Mount slides with Fluoromount-G and seal with a cover slide (keep tissue hydrated throughout this process).
7. Store the slides in a black plastic box at 4 °C to prevent quenching of fluorescence signals until imaging is completed.

3.7 Osteocyte Cx43 HC Activity Imaging and Quantification

1. Begin imaging, using an optical fluorescence microscope, by first standardizing image capture, starting with the slides with the cells expected to produce the brightest signals (the group with the expected highest Cx HC activity) and fix exposure time for comparisons between groups (*see* **Note 1**).
2. Capture images by identifying osteocytes in the cortical bone area located at ~37% of the tibial diaphysis.
3. Using the NIH ImageJ software (*Fiji if preferred*) begin the analysis by preparing images for analysis by first succumbing images to batch conversion into appropriate file format and enhancement, if needed (*see* **Note 15**).
4. Using this same ImageJ software, quantify the fluorescence intensity of EB within osteocytes making sure to do a minimum of 20 cells per section (three sections per mouse) (*see* **Note 16**).
5. Once all images have been analyzed, an ANOVA/t-test can be used to determine statistical significance (Fig. 6).

4 Notes

1. Cx HCs are permeable to Evans blue (EB) (Mw 960.81 Da), which possesses four negative charges and it is a fluorescent dye with two excitation wavelengths (λex) at 470 nm and 540 nm and a peak emission (λem) at 680 nm, producing fluorescence in the red to far-red spectrum. More importantly, this molecule can be fixed by fixative reagent to allow the detection on the tissue sections [22]. Because of low cell permeability to EB, it has been used to evaluate cell viability. EB can bind to albumin protein, so it has also been used to tract the integrity of blood barriers such as the blood–brain barrier [23]. It has also been reported that EB blocks the activity of glutamate receptors and the transport and activity of P2 receptors [24–26]. Pannexin 1 are high-capacity channels that share similar permeability and pharmacology with Cx HCs. However, it has been shown that they are impermeable to EB [13]. Of note, other small fluorescence dyes can be used, however ensure the dyes can be fixed on tissue sections, prior to adoption for this application.

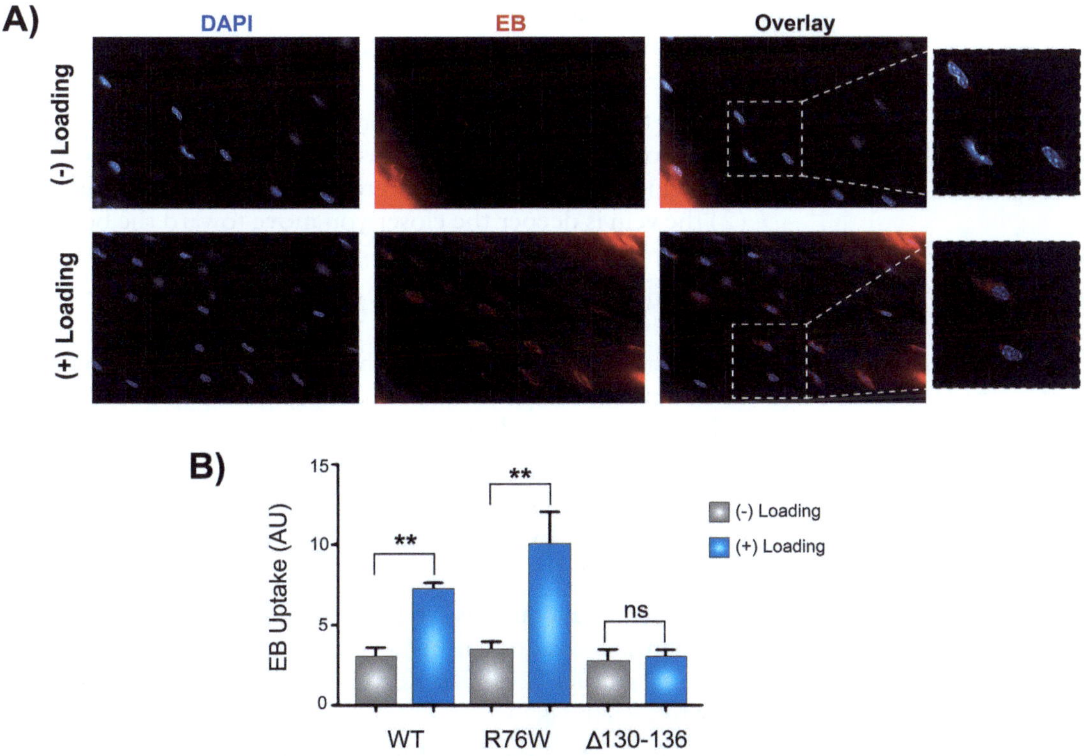

Fig. 6 Imaging and quantification of in vivo osteocyte Cx43 activity. (**a**) Imaging of mice with osteocyte Cx43 HCs activated ((+) Loading) and nonactivated ((−) Loading) in the cortical bone in the tibial diaphysis. Callouts, highlighting the ability to visualize EB uptake (red) in the cytoplasm of osteocytes with activated/functional HCs. Only WT mice shown. (**b**) Quantification of EB uptake, demonstrating the level of permeable osteocyte Cx43 HCs, synonymous to HC activity. Here, WT and mutant R76W mice have functional osteocytic Cx43 HCs, while the mutant Δ130–136 mice lack active HCs, thus see no significant increase of EB uptake [21]

2. This protocol of the relationship between applied compressive loading and bone tissue deformation of the left tibia was established for 15-week-old mice in vivo based on previously reported protocols [27, 28].
3. For our experiments, we typically only calculate these metrics/load mice unilaterally, meaning for us the left tibia is measured/loaded and the right tibia is used as a contralateral, nonloaded control.
4. We typically select peak forces to generate peak periosteal strains of 1200 με at the cortical midshaft since this strain level has been shown to elicit an anabolic response at this region previously [29].
5. Since warming the mouse/tail causes vasodilation, this technique can be utilized for better vein visibility. This can be done using a heating chamber, heating pad, a heating lamp, or dipping the tail in warm water, prior to injection. However, caution should be taken to avoid overheating animal.

6. The mouse has two lateral tail veins and a tail artery located on the ventrum of the tail, either the left or right lateral tail veins may be used.
7. During this step, it is the best practice to do injections close to the tip of the tail since (1) in case first injection is unsuccessful and vein is perforated, you can retry by advancing upward, and (2) the vein is deeper the closer you move toward the body and can be more difficult to access.
8. If injection is successful, blood should flash to the hub of the needle, vein will lighten, and liquid will flow easily during administration. No resistance should be felt. Visibly, mice will also begin to change to a blueish color, the snout, paws, and tail will turn blue, indicating successful administration within ~5 min.
9. A clotting agent, such as styptic powder, silver nitrate sticks, or cautery pen, can be used to assist with the stoppage of blood flow.
10. The cyclic axial compressive load being applied to the left tibia of each mouse is done using a custom loading device based on previous studies [27, 28]. Keep in mind 1 gram (g) = 0.0098 Newton earth (N).
11. Blunt needed is optional, but preferred, to reduce the risk of puncturing the aorta. Blunting of the needle can be done manually using an assortment of tools/sandpaper. Additionally, needle of bigger gauges/not perfusions based, at the preference of the investigator.
12. If blunt needle is used and there is resistance when trying to insert needle into the left ventricle, prior to insertion, a very small incision can be made to simplify needle entrance. It is recommended before injection, the midline and apex of the heart be identified to ensure accurate insertion of needle into left ventricle.
13. For more detail protocol on perfusion of rodents please see [30, 31].
14. Take special care to not immerse samples too quickly to avoid "cracking" of tissue samples.
15. Though image conversion/enhancement can be done (example see: [32]) ensure images are not manipulated ununiformly skewing results.
16. For a more in detail example of image processing/quantification please see [32].

Acknowledgments

This work was supported by NIH CA148724 & NIH F32DK134051 (to F.M.A.), AR072020 (to J.X.J.) and Welch Foundation grant: AQ-1507 (to J.X.J.)
Figures partially created with BioRender.com.

References

1. Laird DW, Lampe PD (2018) Therapeutic strategies targeting connexins. Nat Rev Drug Discov 17(12):905–921. https://doi.org/10.1038/nrd.2018.138
2. Goodenough DA, Paul DL (2003) Beyond the gap: functions of unpaired connexon channels. Nat Rev Mol Cell Biol 4(4):285–294. https://doi.org/10.1038/nrm1072
3. Sáez JC, Leybaert L (2014) Hunting for connexin hemichannels. FEBS Lett 588(8):1205–1211. https://doi.org/10.1016/j.febslet.2014.03.004
4. Cherian PP, Siller-Jackson AJ, Gu S, Wang X, Bonewald LF, Sprague E, Jiang JX (2005) Mechanical strain opens connexin 43 hemichannels in osteocytes: a novel mechanism for the release of prostaglandin. Mol Biol Cell 16(7):3100–3106. https://doi.org/10.1091/mbc.e04-10-0912
5. Retamal MA, Fernandez-Olivares A, Stehberg J (2021) Over-activated hemichannels: a possible therapeutic target for human diseases. Biochim Biophys Acta Mol Basis Dis 1867(11):166232. https://doi.org/10.1016/j.bbadis.2021.166232
6. Puebla C, Retamal MA, Acuña R, Sáez JC (2017) Regulation of connexin-based channels by fatty acids. Front Physiol 8:11. https://doi.org/10.3389/fphys.2017.00011
7. Pogoda K, Kameritsch P, Retamal MA, Vega JL (2016) Regulation of gap junction channels and hemichannels by phosphorylation and redox changes: a revision. BMC Cell Biol 17 (Suppl 1):11. https://doi.org/10.1186/s12860-016-0099-3
8. Quan Y, Du Y, Wu C, Gu S, Jiang JX (2021) Data on connexin 43 hemichannels regulation of cellular redox in lens. Data Brief 39:107572. https://doi.org/10.1016/j.dib.2021.107572
9. Riquelme MA, Cardenas ER, Xu H, Jiang JX (2020) The role of connexin channels in the response of mechanical loading and unloading of bone. Int J Mol Sci 21(3). https://doi.org/10.3390/ijms21031146
10. Schalper KA, Riquelme MA, Brañes MC, Martínez AD, Vega JL, Berthoud VM, Bennett MV, Sáez JC (2012) Modulation of gap junction channels and hemichannels by growth factors. Mol BioSyst 8(3):685–698. https://doi.org/10.1039/c1mb05294b
11. Hu Z, Riquelme MA, Gu S, Jiang JX (2020) Regulation of connexin gap junctions and hemichannels by calcium and calcium binding protein calmodulin. Int J Mol Sci 21(21). https://doi.org/10.3390/ijms21218194
12. Gaete PS, Contreras JE (2021) A method for assessing ionic and molecular permeation in connexin hemichannels. Methods Enzymol 654:271–293. https://doi.org/10.1016/bs.mie.2021.01.009
13. Cea LA, Cisterna BA, Puebla C, Frank M, Figueroa XF, Cardozo C, Willecke K, Latorre R, Sáez JC (2013) De novo expression of connexin hemichannels in denervated fast skeletal muscles leads to atrophy. Proc Natl Acad Sci USA 110(40):16229–16234. https://doi.org/10.1073/pnas.1312331110
14. Johnson RG, Le HC, Evenson K, Loberg SW, Myslajek TM, Prabhu A, Manley AM, O'Shea C, Grunenwald H, Haddican M, Fitzgerald PM, Robinson T, Cisterna BA, Sáez JC, Liu TF, Laird DW, Sheridan JD (2016) Connexin hemichannels: methods for dye uptake and leakage. J Membr Biol 249(6):713–741. https://doi.org/10.1007/s00232-016-9925-y
15. Orellana JA, Díaz E, Schalper KA, Vargas AA, Bennett MV, Sáez JC (2011) Cation permeation through connexin 43 hemichannels is cooperative, competitive and saturable with parameters depending on the permeant species. Biochem Biophys Res Commun 409(4):603–609. https://doi.org/10.1016/j.bbrc.2011.05.031
16. Riquelme MA, Gu S, Hua R, Jiang JX (2021) Mechanotransduction via the coordinated actions of integrins, PI3K signaling and connexin hemichannels. Bone Res 9(1):8. https://doi.org/10.1038/s41413-020-00126-w
17. Zhang C, Yan Z, Maknojia A, Riquelme MA, Gu S, Booher G, Wallace DJ, Bartanusz V, Goswami A, Xiong W, Zhang N, Mader MJ, An Z, Sayre NL, Jiang JX (2021) Inhibition of

astrocyte hemichannel improves recovery from spinal cord injury. JCI Insight 6(5). https://doi.org/10.1172/jci.insight.134611

18. Zhao D, Riquelme MA, Guda T, Tu C, Xu H, Gu S, Jiang JX (2022) Connexin hemichannels with prostaglandin release in anabolic function of bone to mechanical loading. elife 11. https://doi.org/10.7554/elife.74365
19. Batra N, Kar R, Jiang JX (2012) Gap junctions and hemichannels in signal transmission, function and development of bone. Biochim Biophys Acta 1818(8):1909–1918. https://doi.org/10.1016/j.bbamem.2011.09.018
20. Loiselle AE, Jiang JX, Donahue HJ (2013) Gap junction and hemichannel functions in osteocytes. Bone 54(2):205–212. https://doi.org/10.1016/j.bone.2012.08.132
21. Xu H, Gu S, Riquelme MA, Burra S, Callaway D, Cheng H, Guda T, Schmitz J, Fajardo RJ, Werner SL, Zhao H, Shang P, Johnson ML, Bonewald LF, Jiang JX (2015) Connexin 43 channels are essential for normal bone structure and osteocyte viability. J Bone Miner Res Off J Am Soc Bone Miner Res 30(3):436–448. https://doi.org/10.1002/jbmr.2374
22. Hed J, Dahlgren C, Rundquist I (1983) A simple fluorescence technique to stain the plasma membrane of human neutrophils. Histochemistry 79(1):105–110. https://doi.org/10.1007/bf00494347
23. Honeycutt SE, O'Brien LL (2021) Injection of Evans blue dye to fluorescently label and image intact vasculature. BioTechniques 70(3):181–185. https://doi.org/10.2144/btn-2020-0152
24. Schürmann B, Wu X, Dietzel ID, Lessmann V (1997) Differential modulation of AMPA receptor mediated currents by Evans blue in postnatal rat hippocampal neurones. Br J Pharmacol 121(2):237–247. https://doi.org/10.1038/sj.bjp.0701125
25. Price CJ, Raymond LA (1996) Evans blue antagonizes both alpha-amino-3-hydroxy-5-methyl-4-isoxazolepropionate and kainate receptors and modulates receptor desensitization. Mol Pharmacol 50(6):1665–1671
26. Wittenburg H, Bültmann R, Pause B, Ganter C, Kurz G, Starke K (1996) P2-purinoceptor antagonists: II. Blockade of P2-purinoceptor subtypes and ecto-nucleotidases by compounds related to Evans blue and trypan blue. Naunyn Schmiedeberg's Arch Pharmacol 354(4):491–497. https://doi.org/10.1007/bf00168441
27. Lynch ME, Main RP, Xu Q, Walsh DJ, Schaffler MB, Wright TM, van der Meulen MC (2010) Cancellous bone adaptation to tibial compression is not sex dependent in growing mice. J Appl Physiol (1985) 109(3):685–691. https://doi.org/10.1152/japplphysiol.00210.2010
28. De Souza RL, Matsuura M, Eckstein F, Rawlinson SC, Lanyon LE, Pitsillides AA (2005) Non-invasive axial loading of mouse tibiae increases cortical bone formation and modifies trabecular organization: a new model to study cortical and cancellous compartments in a single loaded element. Bone 37(6):810–818. https://doi.org/10.1016/j.bone.2005.07.022
29. Sugiyama T, Price JS, Lanyon LE (2010) Functional adaptation to mechanical loading in both cortical and cancellous bone is controlled locally and is confined to the loaded bones. Bone 46(2):314–321. https://doi.org/10.1016/j.bone.2009.08.054
30. Gage GJ, Kipke DR, Shain W (2012) Whole animal perfusion fixation for rodents. J Vis Exp 65. https://doi.org/10.3791/3564
31. Wu J, Cai Y, Wu X, Ying Y, Tai Y, He M (2021) Transcardiac perfusion of the mouse for brain tissue dissection and fixation. Bio-protocol 11(1):e3876. https://doi.org/10.21769/BioProtoc.3876
32. Potter JA, Price GW, Cliff CL, Williams BM, Hills CE, Squires PE (2021) Carboxyfluorescein dye uptake to measure connexin-mediated hemichannel activity in cultured cells. Bio-protocol 11(3):e3901. https://doi.org/10.21769/BioProtoc.3901

Chapter 10

Protocol for the Study of Connexin and DNA Interactions

Ainoa Fernandez-Olivares and Mauricio A. Retamal

Abstract

Connexins (Cxs) are transmembrane proteins which form hemichannels and gap junction channels at the plasma membrane. These channels allow the exchange of ions and molecules between the intra- and extracellular space and between cytoplasm of adjacent cells, respectively. The channel function of Cx assemblies has been extensively studied; however, "noncanonical" functions have emerged in the last few decades and have capture the attentions of many researchers, including the role of some Cxs as gene modulators or transcription factors. In this chapter, we describe a protocol to study the interaction of Cx46 with DNA in HeLa cells. These methods can facilitate understanding the role of Cxs in physiological processes and pathological mechanisms, including, for example, the contribution of Cx46 in maintaining stemness of glioma cancer stem cells.

Key words Connexins, Chromatin immunoprecipitation (Chip), DNA interactions, Immunoprecipitation, Cx46

1 Introduction

Connexins (Cxs) are a family of proteins that usually are present at the plasma membrane forming hemichannels or gap junction channels [1]. Although hemichannels and gap junction channels are formed by the same types of proteins, they participate in different cellular functions. Thus, while hemichannels allow the exchange of molecules and ions between the cytoplasm and the extracellular milieu, gap junction channels allow the exchange of molecules and ions directly from a cytoplasm on one cell to the cytoplasm and neighboring cells [2, 3]. The channel-dependent way of action is known as "canonical" function, and it has been extensively studied over many years. However, also a "noncanonical" role of Cxs has emerged in the last years [4]. A very interesting mechanisms of action is related to the capacity of Cx43 and Cx46 to be localized in the cell nucleus and modulate gene expression [5–7]. However, the molecular mechanisms underlying these phenomena, remain elusive.

Fabio Mammano and Mauricio Retamal (eds.), *Connexin Hemichannels: Methods and Protocols*, Methods in Molecular Biology, vol. 2801, https://doi.org/10.1007/978-1-0716-3842-2_10,

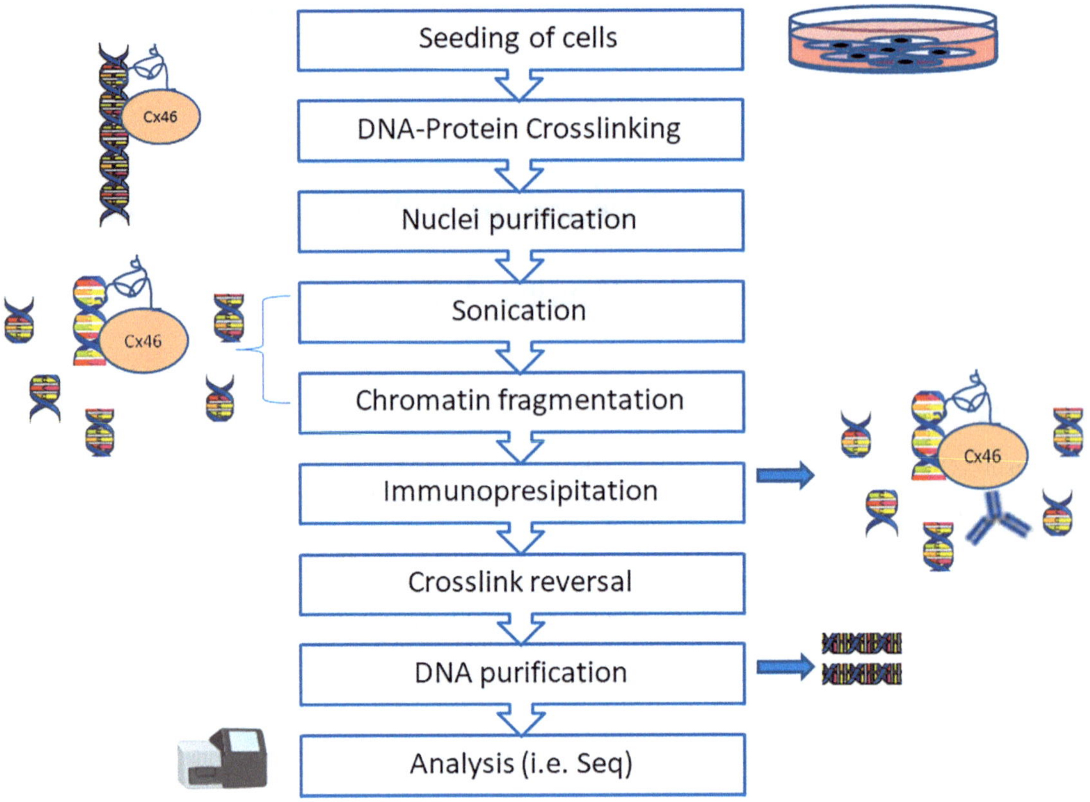

Fig. 1 Scheme of chromatin immunoprecipitation (Chip) protocol. Briefly, cells are grown in a plastic dish, then proteins that are interacting with the DNA are crosslinked, to preserve these interactions. Cell nuclei are purified and sonicated to disrupt their content and to break down the DNA in small fragments. Subsequently, an antibody against Cx46 is added to the fragmented DNA, incubated and precipitated to obtain only DNA fragments that interact with Cx46. Finally, Cx46 and DNA crosslink is reversed, and free DNA fragments can be analyzed by different techniques, as for example, sequencing

Here, we present a protocol for cell transfection, Cx46–DNA interaction, and analysis of Cx46–DNA interaction (*see* Fig. 1). This protocol will be helpful for future experiments in the field of Cx-based gene regulation in both cellular physiology and pathology.

2 Materials

2.1 Cell Culture

1. HeLa cells.
2. Transmitted light microscope.
3. Bench cooled-centrifuge.
4. 60 mm cell culture plates.
5. Incubator maintained at 37 °C and 5% CO_2.

6. Borosilicate glass Pasteur pipettes.
7. Cell culture medium: Dulbecco's Modified Eagle Medium (DMEM) low glucose, supplemented with 10% fetal bovine serum (FBS) and penicillin (1000 U/mL)/streptomycin (1000 U/mL) (complete DMEM). Stored at 4 °C, warmed before use.
8. 100 mL autoclaved glass bottle.
9. Sterile biosafety cabinet.
10. Trypsin/EDTA stock solution (10×, 0.5% trypsin) is stored at −20 °C. The working (1×) solution is stored at 4 °C and warmed before use.

2.2 HeLa Cells Transfection

1. Transmitted light microscope.
2. 35 mm cell culture plates.
3. pCAG–Cx46 mammalian expression plasmid.
4. Cooled bench centrifuge.
5. Incubator maintained at 37 °C and 5% CO_2.
6. Vacuum line.
7. Complete DMEM, stored at 4 °C.
8. Lipofectamine™ 3000 Reagent, stored at 4 °C.
9. Phosphate-buffered saline (PBS) (sodium chloride 0.137 M, potassium chloride 0.027 M, sodium phosphate dibasic 0.01 M, potassium phosphate monobasic 0.018 M, pH 7.4), stored at 4 °C.
10. Opti-Minimal Essential Medium (OptiMEM), stored at 4 °C.
11. Sterile biosafety cabinet.
12. Blasticidine–HCl selection antibiotic.

2.3 Nuclear Extracts and DNA Fragmentation

1. 100 mm cell culture plates.
2. 37% formaldehyde stock.
3. Orbital shaker.
4. 1.25 mM glycine stock solution.
5. Cell scraper.
6. 15 mL Falcon tube.
7. Cooled bench centrifuge.
8. Lysis buffer (50 mM HEPES pH: 7.8, 3 mM $MgCl_2$, 20 mM KCl, 0.1% NP-40, 2 mM phenylmethylsulfonyl fluoride (PMSF), 2 μg/mL Aprotinin, 2 μg/mL Leupeptin, and 1 μM Pepstatin).
9. Tripan blue.
10. Basket with ice.

11. Tight Dounce homogenizer.
12. 1.5 mL Eppendorf tubes.
13. Sonication buffer (50 mM HEPES pH: 7.8, 140 mM NaCl, 1 mM EDTA, 1% Triton X-100, 0.1% sodium deoxycholate, 1% sodium dodecyl sulfate, 2 mM PMSF, 2 μg/mL Aprotinin, 2 μg/mL Leupeptin, and 1 μM Pepstatin).
14. Bioruptor sonication tubes (Diagenode).
15. Neubauer chamber.
16. Bioruptor Pico sonication device.

2.4 Agarose Gel

1. Agarose.
2. TAE buffer (40 mM Tris–HCl base, 20 mM acetic acid, and 1 mM EDTA).
3. DNA samples.
4. UV transilluminator.
5. SYBR safe nucleic acid gel stain.
6. 1 kb DNA ladder.
7. Any commercial 6× DNA loading buffer.

2.5 Chromatin Immunoprecipitation (Chip)

1. Fractioned DNA samples.
2. Sonication buffer without sodium dodecyl sulfate.
3. Mouse IgG linked to protein G.
4. Orbital shaker.
5. Cooled bench centrifuge.
6. M-280 Sheep Anti-Mouse IgG dynabeads.
7. Anti-connexin 46 antibody (Santa Cruz Biotechnology, C-3).
8. Magjet rack 2 × 1.5 mL (Thermo Scientific).
9. Immunoprecipitation (IP) buffer (100 mM Tris–HCl pH: 8.0, 500 mM LiCl, 1% NP-40, and 0.1% sodium deoxycholate).
10. TE buffer (10 mM Tris–HCl and 1 mM EDTA), pH: 8.0.
11. Elution buffer (50 mM NaHCO3, 1% SDS).
12. Refrigerator at −20 °C.

2.6 Decrosslinking and Purification of DNA

1. 5 M NaCl.
2. 10 mg/mL RNase stock solution.
3. Thermoregulated bath.
4. 50 mg/mL proteinase K solution.
5. TE buffer, pH 8.0.
6. Phenol:chloroform:isoamyl alcohol (1:1:1).
7. Chloroform:isoamyl alcohol (24:1).

8. Cooled bench centrifuge.
9. 1.5 mL Eppendorf tubes.
10. Absolute ethanol.
11. 3 M sodium acetate, pH 7.0.
12. 20 mg/mL glycogen solution.
13. Refrigerator at −20 °C.
14. Fluorimeter for DNA quantification.

3 Methods

3.1 Culturing HeLa Cells

1. Warm complete DMEM in a 37 °C water bath for 15 min prior to use.
2. Culture HeLa cells in sterile 60 mm tissue culture dishes that are housed in an incubator maintained at 37 °C and 5% CO_2 (*see* **Note 1**).
3. Passage HeLa cells once they reach a ~90% confluence (*see* **Note 2**).
4. For subculturing, 60 mm tissue culture dishes are removed from the incubator. Check cell viability using a light microscope and tripan blue exclusion.
5. Attach an autoclaved borosilicate glass Pasteur pipette to a vacuum line and aspirate the growth medium from the dish.
6. Gently rinse the cell monolayer with 1–2 mL of sterile PBS. Use the Pasteur pipette to aspirate PBS between washes.
7. Add 1 mL of trypsin/EDTA 0.05% and place the dish in the incubator for 3 min. Remove the dish from the incubator and check whether cells have detached using a light-transmitted microscope.
8. Once cells are detached, add 1 mL of complete DMEM to inhibit the trypsin.
9. Remove cells using a 1000 mL pipette and place the cell suspension in a 2 mL Eppendorf tube.
10. Centrifuge the tube at ~200 g-force for 5 min at room temperature.
11. Remove trypsin and add 1 mL of complete DMEM. Gently resuspend the cell pellet by pipetting up and down to break apart any cell aggregates.
12. Count the cell number using a Neubauer chamber.
13. Calculate the volume required for seeding 1×10^5 cells and add this volume to a new 60 mm tissue culture dishes.
14. Add 4 mL of thawed complete DMEM.

3.2 HeLa Cells Transfection

1. Warm complete DMEM and OptiMEM in a 37 °C water bath.
2. Check that cell confluence is below 60% using a light microscope.
3. Label two 1.5 mL microcentrifuge tubes for each transfection reaction.
4. Thaw aliquots of plasmid pCAG–Cx46 at room temperature, and place in biosafety cabinet along with Lipofectamine 3000.
5. For each transfection reaction, prepare a tube containing 117.5 μL OptiMEM and 7.5 μL Lipofectamine™ 3000 Reagent. Prepare another tube containing a volume equivalent to 1 μg plasmid DNA and 2 μL of P3000 reagent in a final volume of 125 μL of OptiMEM.
6. After 5 min, mix both tubes by gently pipetting.
7. Incubate the tube with the mixture for 5 min at room temperature.
8. Remove the 35 mm dishes containing HeLa cells and place them in the safety chamber.
9. Add the transfection solution drop-by-drop onto the cells throughout the 35 mm cell culture dish containing 2 mL of complete DMEM.
10. Return the cells to the incubator and incubate overnight.
11. Wash twice with PBS and replace the media-containing plasmid with complete DMEM.
12. After 2 weeks, it is possible to select stable transfected cells by adding 10 μg/mL of blasticidine to the complete DMEM.
13. If cells reach more than 90% confluence, repeat **steps** from **6** to **11** adding 10 μg/mL of blasticidine.

3.3 Nuclear Extracts

1. Rinse twice the cell cultured in 100 mm plates (at ~80% confluence) with 5 mL of PBS at room temperature.
2. Add 5 mL of formaldehyde at 37% in PBS to bring it to a final concentration of 1% for each plate and perform crosslinking for 10 min in orbital agitation (*see* **Note 3**).
3. Add 0.5 mL of 1.25 mM glycine solution and incubate for 5 min at room temperature and constant orbital agitation.
4. Discharge fixing solution and rinse with 5 mL of PBS × three times (*see* **Note 4**).
5. Discharge the last PBS rinse and harvest cells using 5 mL of PBS containing protease inhibitors.
6. Place the cell suspension in a 15 mL Falcon tube and centrifuge for 5 min at 2000 g-force and 4 °C.

7. Resuspend the pellet in 1.5 mL lysis buffer and count the cells with trypan blue to obtain the total number of cells.
8. Incubate the cell suspension on ice for 10 min.
9. Homogenize the cell suspension with a 2 mL Tight Dounce homogenizer by applying ~10 strokes. Check with trypan blue if the nuclei have been purified.
10. Transfer the homogenized to a 1.5 mL Eppendorf tube.
11. Centrifuge for 5 min at 500 g-force and 4 °C, and discharge supernatant.
12. Resuspend the pellet in 2 mL of sonication buffer and transfer it to a Bioruptor sonication tubes (Diagenode). Add a maximum volume of 500 μL of resuspended nuclei.
13. Sonicate for 50 min at 30 s intervals using a Bioruptor Pico sonication device. Check for rupture of the nuclei using trypan blue in a Neubauer chamber.
14. Centrifuge at 10,000 g-force for 5 min at 4 °C.
15. Transfer the supernatant to a new 1.5 mL Eppendorf.
16. To assess the degree of chromatin fragmentation, perform rapid decrosslinking of an aliquot in a thermocycler at 95 °C for 25 min, then add 2 μg/mL of RNase A and incubat at 37 °C for 15 min. Finally, add 50 μg/mL of proteinase K and incubate for 15 min at 58 °C.
17. Quantify uncrosslinked fragmented chromatin using the Qubit dsDNA BR Assay Kit on a Qubit Fluorometer following the manufacturer's instructions.

3.4 Agarose Gel

1. Prepare a 1% agarose gel in TAE buffer.
2. Mix 100 ng of each DNA sample, 3 μL of 6× loading buffer and enough H_2O to a final volume of 18 μL.
3. Load this mixture in the agarose gel.
4. Run gel at 100 V until the sample buffer is located at approximately 50% of the gel.
5. Place the gel in a UV transilluminator and examine the level of chromatin fragmentation, ideally, it should be between 200 and 300 bp (*see* **Note 5**).

3.5 Chromatin Immunoprecipitation Protocol

1. Take 200 ng of DNA from nonimmunoprecipitated samples and keep it as control (Input) at −20 °C.
2. To 400 ng of fragmented DNA (from point Subheading 3.3, **step 15**) add enough sonication buffer without sodium dodecyl sulfate to complete a final volume of 500 μL.
3. Add 2 μL of mouse IgG (12–371, Merck Millipore) linked to protein G in the tube containing fragmented chromatin.

4. Incubate this mixture for 2 h under orbital agitation at 4 °C.
5. Centrifuge at 2000 g-force for 5 min at 4 °C and transfer the supernatant to a new 1.5 mL tube.
6. Add 50 μL of M-280 Sheep Anti-Mouse IgG dynabeads and 2 μg of anti-connexin 46 antibody (Santa Cruz Biotechnology, C-3).
7. Incubate overnight under constant rotation at 4 °C.
8. Place the tube over a magnet until all the beads are concentrated at the bottom.
9. Discharge the supernatant and rinse the pellet with 500 μL of IP wash buffer for 5 min in orbital shaking.
10. Repeat **steps 8** and **9** once more.
11. Discharge the supernatant and rinse the beads with 500 μL of TE buffer, pH: 8.0, for 5 min in orbital shaking.
12. Repeat **steps 8** and **9** once more.
13. In the last rinse, remove the supernatant and resuspend the in 100 μL of elution buffer. Place in in a vortex mixer and run at 2800 g-force.
14. Incubate samples 15 min at 65 °C in a thermoregulated bath, then store the supernatant at −20 °C.

3.6 Decrosslinking and Purification of DNA

1. Add the necessary volume of elution buffer to the control samples (Inputs) to reach a final volume 100 μL (*see* **Note 6**).
2. Add 4 μL of 5 M NaCl to the immunoprecipitated what and to the input tubes to obtain a final concentration of 200 mM.
3. Add 2 μg/mL of RNase A and incubate overnight in a thermoregulated bath at 65 °C.
4. Add 50 μg/mL of proteinase K and incubate in a thermoregulated bath for 2 h at 50 °C.
5. Add 200 μL of TE buffer (pH 8.0) and 300 μL of phenol: chloroform:isoamyl alcohol (1:1:1).
6. Invert the tubes upside down and then returns it to its normal position 4–6 times.
7. Centrifuge at 15,000 g-force for 10 min at 4 °C.
8. Recover the aqueous phase and place it in a 1.5 mL Eppendorf tube.
9. Add 300 mL of chloroform:isoamyl alcohol (24:1) and invert the tube upside down and then return it to its normal position 4–6 times.
10. Centrifuge at 12,000 g-force for 10 min at 4 °C.

11. Recover the aqueous phase and place it in a 1.5 mL Eppendorf tube. Add 2.5 volumes of absolute ethanol, 0.1 volumes of 3 M sodium acetate pH 7.0, and 20 μg of glycogen.
12. Place the mixture at −20 °C overnight.
13. Centrifuge at 7500 g-force for 25 min at 4 °C.
14. Rinse the pellet with 500 μL of 70% ethanol.
15. Centrifuge again for 10 min at 7500 g-force for 25 min at 4 °C (*see* **Note 7**).
16. Discharge supernatant and resuspend the pellet in 30 μL of TE buffer or 100 μL in the case of inputs.
17. Measure DNA concentration using the Qubit HS dsDNA Assay Kit (*see* **Note 8**).
18. The DNA samples are now ready for qPCR or sequencing.

4 Notes

1. This protocol was adjusted for HeLa cells. However, it can be applied also to other cell lines.
2. It is important not let the cell grow beyond 90% confluence, to avoid cell stress, which can alter expected results.
3. If the type of interactions between a given Cx and the DNA is unknown, different crosslinker can be used with different spacer sizes, like EGS (ethylene glycol bis (sulfosuccinimidyl succinate).
4. The fixing solution must be discarded in a proper vessel.
5. The size of fragmented DNA could be more than 300 bp but never more than 500 bp.
6. Control samples are fragmented but not immunoprecipitated DNA.
7. Let the pellet dry but not too much, because otherwise resuspension turns problematic.
8. Dilute the DNA 1:10 in water because the quantification is more reliable than in TE.

Acknowledgments

The authors thank to FONDECYT grants #1160227.

References

1. Sáez JC, Berthoud VM, Brañes MC et al (2003) Plasma membrane channels formed by connexins: their regulation and functions. Physiol Rev 83:1359–1400. https://doi.org/10.1152/physrev.00007.2003
2. Saez JC, Connor JA, Spray DC, Bennett MVL (1989) Hepatocyte gap junctions are permeable to the second messenger, inositol 1,4,5-trisphosphate, and to calcium ions. Proc Natl Acad Sci USA 86:2708–2712. https://doi.org/10.1073/pnas.86.8.2708
3. Retamal MA, Reyes EP, García IE et al (2015) Diseases associated with leaky hemichannels. Front Cell Neurosci 9. https://doi.org/10.3389/fncel.2015.00267
4. Van Campenhout R, Cooreman A, Leroy K et al (2020) Non-canonical roles of connexins. Prog Biophys Mol Biol. https://doi.org/10.1016/j.pbiomolbio.2020.03.002
5. Epifantseva I, Xiao S, Baum RE et al (2020) An alternatively translated connexin 43 isoform, GJA1-11k, localizes to the nucleus and can inhibit cell cycle progression. Biomol Ther 10:473. https://doi.org/10.3390/biom10030473
6. Kotini M, Barriga EH, Leslie J et al (2018) Gap junction protein connexin-43 is a direct transcriptional regulator of N-cadherin in vivo. Nat Commun 9:3846. https://doi.org/10.1038/s41467-018-06368-x
7. Vitale ML, Garcia CJ, Akpovi CD, Pelletier R-M (2017) Distinctive actions of connexin 46 and connexin 50 in anterior pituitary folliculostellate cells. PLoS One 12:e0182495. https://doi.org/10.1371/journal.pone.0182495

Chapter 11

Enhanced Methodologies for Investigating the Electrophysiological Characteristics of Endogenous Pannexin 1 Intercellular Cell–Cell Channels

Nicolás Palacios-Prado, Teresa Vergara, and Juan C. Sáez

Abstract

Gap junctions, pivotal intercellular conduits, serve as communication channels between adjacent cells, playing a critical role in modulating membrane potential distribution across cellular networks. The family of Pannexin (Panx) proteins, in particular Pannexin1 (Panx1), are widely expressed in vertebrate cells and exhibit sequence homology with innexins, the invertebrate gap junction channel constituents. Despite being ubiquitously expressed, detailed functional and pharmacological properties of Panx1 intercellular cell–cell channels require further investigation. In this chapter, we introduce optimized cell culture methodologies and electrophysiology protocols to expedite the exploration of endogenous Panx1 cell–cell channels in TC620 cells, a human oligodendroglioma cell line that naturally expresses Panx1. We anticipate these refined protocols will significantly contribute to future characterizations of Panx1-based intercellular cell–cell channels across diverse cell types and offer valuable insights into both normal cellular physiology and pathophysiology.

Key words Pannexins, Intercellular channels, junctional conductance, Panx1

1 Introduction

Intercellular channels facilitate cytoplasmic communication among neighboring cells in most vertebrate and invertebrate organisms. These channels play a critical role in regulating the bioelectric potential distribution within cellular networks and tissue organization [1]. This, in turn, influence a host of cellular activities, encompassing transcription, proliferation, cellular architecture, metabolic and electrical synchronization, migration, and differentiation [2, 3].

A cluster of intercellular channels collectively forms a gap junction plaque, an organized crystalline structure. In vertebrates, these plaques are formed by hexameric oligomers of connexin proteins. Each hexamer forms a hemichannel, with the junction of two

Fabio Mammano and Mauricio Retamal (eds.), *Connexin Hemichannels: Methods and Protocols*, Methods in Molecular Biology, vol. 2801, https://doi.org/10.1007/978-1-0716-3842-2_11,

hemichannels from contacting cells forming a complete gap junction channel. This facilitates intercellular transfer of ions, metabolites, and signaling molecules [4]. In a similar manner, pannexin 1 (Panx1) proteins oligomerize into heptamers [5] to form a Panx channel (pannexon) in the plasma membrane. This channel allows the passage of chloride ions, metabolites such as glucose, ATP, and other signaling molecules upon various cellular stimuli [6, 7]. Both endogenous and exogenous Panx1 proteins have been demonstrated to establish intercellular channels [8–10]. Despite this, the existence of tightly packed crystalline structures, akin to connexin gap junction plaques, is yet to be confirmed for pannexin-based cell–cell channels [11, 12]. This suggests Panx cell–cell channels likely operate as disperse intercellular conduits in the plasma membrane [10].

Exploring the electrophysiology of endogenous Panx1 cell–cell channels presents a number of challenges: potential co-existence of endogenous connexin-based gap junction channels requiring connexin pharmacological blockers, and the relatively low total junctional conductance ($g_j \sim 1$ nS) and small single cell–cell channel conductance (~35 pS for the S-state [10]) necessitating a low-noise electrophysiological setup for dual whole-cell patch clamp.

Protein expression of Panx1 is observed across various tissues, excluding soft and adipose tissues [13, 14]. Notably, unlike connexin-based gap junction channels, Panx1-mediated electrical coupling only occurs in high confluence cell cultures and exhibits low sensitivity to high concentration (1 mM) of connexin-based gap junction channel blockers, heptanol and octanol [9, 10]. These attributes greatly influence the electrophysiological characterization of Panx1 cell–cell channels.

In this chapter, we provide a detailed protocol for examining the electrophysiological properties of endogenous Panx1 cell–cell channels within an oligodendroglioma-derived cell line (TC620), leveraging the dual whole-cell patch clamp technique.

2 Materials

2.1 Cell Culture

1. TC620 cells.
2. Transmitted light microscope.
3. Table-top centrifuge.
4. Cell culture dishes (60 and 35 mm).
5. CO_2 Incubator (set at 37 °C and 5% CO_2).
6. Sterile borosilicate glass Pasteur pipettes with cotton plug (2 mL).
7. Pipette rubber bulb.
8. Sterile borosilicate glass pipettes (10 mL).

9. Motorized pipette controller.
10. Cell culture medium: Iscove's Modified Dulbecco's Medium (IMDM), supplemented with 10% fetal bovine serum (FBS) and penicillin (1,000 U/mL)/streptomycin (1,000 U/mL), pH = 7.4. Stored at 4 °C, warmed at 37 °C before use.
11. Sterile glass bottle (100 and 1,000 mL).
12. Sterile biosafety cabinet.
13. Sterile phosphate buffer saline, pH = 7.4 (PBS).
14. Trypsin/EDTA stock solution (10×, 0.5% trypsin) is stored at −20 °C. The working (1×) solution is stored at 4 °C and warmed at 37 °C before use.
15. Vacuum line.
16. Erlenmeyer flask (500 mL).
17. Bottle top vacuum filter (500 mL, 0.22 μm).
18. Ethanol (70%).
19. Powder free latex gloves.
20. Single-channel manual micropipette: P1000, P100, P20, P10.
21. Micropipette tips: 1000, 200, and 20 μL.
22. Sterile conical centrifuge tubes (15 mL).
23. Thermoregulated bath (37 °C).
24. pH meter.
25. Sterile glass coverslips (12 mm).
26. Sterile high precision tweezer.

2.2 Cell Counting

1. Cell culture medium: Iscove's Modified Dulbecco's Medium (IMDM), supplemented with 10% fetal bovine serum (FBS) and penicillin (1,000 U/mL)/streptomycin (1,000 U/mL), pH = 7.4. Stored at 4 °C, warmed to 37 °C prior to use.
2. Transmitted light microscope.
3. Centrifuge tube (500 μL).
4. Haemocytometer cell counting chamber.
5. Square cover glass (20 × 20 mm).
6. Trypan blue stain.
7. Sterile biosafety cabinet.
8. Sterile phosphate buffer saline, pH = 7.4 (PBS).
9. Sterile Borosilicate glass Pasteur pipettes with cotton plug (2 mL).
10. Pipette rubber bulb.
11. Handheld tally counter.

2.3 Electrophysiology Medium

1. Sodium chloride.
2. Sodium aspartate.
3. Sodium hydroxide.
4. Potassium chloride.
5. Potassium hydroxide.
6. Magnesium chloride.
7. Hydrochloric acid.
8. Calcium chloride.
9. Cesium chloride.
10. Cesium hydroxide.
11. Barium chloride.
12. Ethyleneglycol- bis(β-aminoethyl)-N,N,N′,N′-tetraacetic acid (EGTA).
13. 4-(2-hydroxyethyl)-1-piperazineethanesulfonic acid (HEPES).
14. Octanol.
15. Mini vortex shaker.
16. pH Meter.
17. Osmometer (optional).

2.4 Electrophysiology

1. Inverted light microscope with a X-Y translation system.
2. Objective lens (40× or 60× magnification).
3. Fixed-stage platform.
4. Stage adapter for recording chamber.
5. Recording chamber for low-noise electrophysiology.
6. Micromanipulator (×2).
7. Macromanipulator (×2).
8. Micropipette puller.
9. Borosilicate capillary glass for patch pipette (0.68 mm inner diameter (ID) and 1.20 mm outer diameter (OD) with filament).
10. Syringe pore filter (0.2 μm).
11. Micro-loader tips (20 μL).
12. Patch pipette holder (×2).
13. Pipette pressure system with three-way stopcock.
14. Gravity-fed bath superfusion system with two separate perfusion reservoirs, three-way stopcock, air trap and flow regulator (thumb-wheel).
15. Temperature controller for perfusion.
16. Vacuum line.

17. Silicone tubing for perfusion system.
18. Precision tweezer.
19. Anti-vibration table.
20. Compressed air tank (for antivibration table).
21. Faraday cage.
22. Electrophysiology amplifier (two axopatch or one Multiclamp 700B; preferred).
23. Data acquisition system (Digitada).
24. PC with electrophysiology software (i.e., P-Clamp).
25. Flat display screen.

3 Methods

3.1 Culturing TC620 Cells

1. Warm complete IMDM in a 37 °C water bath for 15 min prior to use.
2. Culture TC620 cells in sterile 60 mm tissue culture dishes. Keep these dishes in an incubator set at 37 °C with 5% CO_2.
3. Passage TC620 cells when they reach approximately 80–90% confluence (*see* **Note 1**).
4. Connect a sterile borosilicate glass Pasteur pipette to a vacuum line and aspirate the growth medium from the dish.
5. Gently rinse the cell monolayer with 5 mL of sterile PBS. Use the Pasteur pipette to aspirate PBS.
6. Add 2 mL of prewarmed 0.05% trypsin/EDTA at 37 °C and incubate the dish for 2 min. Subsequently, take out the dish from the incubator and check cell detachment under a light-transmitted microscope.
7. Gently pipetting the trypsin/EDTA medium to facilitate cell detachment.
8. When cells are detached, add 3 mL of complete IMDM to inhibit trypsin activity.
9. Collect cells using a 2 mL Pasteur pipette and transfer the cell suspension to a 15 mL conical centrifuge tube.
10. Spin the tube at 700 G-force for 5 min at room temperature.
11. Aspirate all supernatant with a Pasteur pipette connected to a vacuum line, leaving the cell pellet intact at the bottom.
12. Add 2 mL of complete IMDM. Gently resuspend the cell pellet by pipetting with a Pasteur pipette to break apart any cell clusters.

3.2 Cell Counting

1. Add 10 μL of the resuspended cell medium into a 0.5 mL centrifuge tube, then add 10 μL of trypan blue.
2. Dispense 20 μL of the "trypan blue + cells" solution into a Neubauer chamber, topped with a squared coverslip. Place this setup under a light microscope, adjusting the focus to clear view of the cells and grid lines within the chamber.
3. Using a handheld tally counter, enumerate the live cells (*see* **Note 2**) within one of the nine boxes. Record the count for the first box, reset the counter and repeat the process for remaining eight boxes.
4. Calculate the average number of live cells per box. Multiply this number by 20,000 to estimate the total cell count per mL in the original solution (prior to the addition of Trypan Blue).
5. Determine the required volume for seeding 7–8 × 10^4 cells and dispense this volume into a 15 mL conical centrifuge tube. Fill up with IMDM to a final volume of 2 mL per 35 mm culture dish.
6. With a P1000 micropipette, homogenize 2 mL cell solution and transfer it to a new 35 mm tissue culture dishes, each containing three 12 mm sterile glass coverslips.
7. Inspect the 35 mm culture dish under a light microscope to ensure that glass coverslips are not overlapping.
8. Transfer the culture dishes to an incubator for overnight incubation.
9. After a 24 h period, assess the cell confluency using a light microscope by making a qualitative visual estimation.

3.3 Electrophysiology

3.3.1 Solution Preparation

Begin by preparing 500 mL of 1 M stock solutions for NaCl, NaOH, KCl, HCl, CsCl, CsOH, $MgCl_2$, $CaCl_2$, $BaCl_2$, and HEPES. Store these solutions in 500 mL glass bottles at 4 °C.

Bath Solution

The bath solution composition (in millimolar) includes NaCl, 140; KCl, 5; $CaCl_2$, 1.8; $MgCl_2$, 1; $BaCl_2$, 2; and HEPES, 5. Prepare 1L of bath solution using the stock solutions, adjusting pH to 7.4 with HCl and osmolarity to 300 mOsml/dm^3 (+/− 10 mOsml/dm^3). Filter the solution through a 0.22 μm bottle top vacuum filter into a sterile 1 L glass bottle for storage.

Octanol Solution

Add 50 mL of bath solution into a 50 mL conical centrifuge tube. To attain a final concentration of 1 mM, add 7.8 μL of octanol (density 0.827 g/mL) to the tube and mix thoroughly for 2 min using a mini vortex shaker.

Pipette Solution

The pipette solution (in millimolar) comprises CsCl, 130; NaAsp, 10; $MgCl_2$, 1; $CaCl_2$, 0.2; EGTA, 2; and HEPES, 5. Prepare a 100 mL pipette solution using stock solutions, then add the appropriate quantity of NaAsp and EGTA powder. Adjust the pH to 7.2 with CsOH and the osmolarity at 290 mOsml/dm^3 (+/− 10 mOsml/dm^3). Aliquot the pipette solution into 2 mL centrifuge tubes and store them at −20 °C. Prior to recording, thaw one aliquot, transfer the solution to a 5 mL syringe equipped with a 0.2 μm syringe disc filter and a microloader tip for loading the recording pipettes.

3.3.2 Whole Cell Patch Clamp Procedure

Patch clamp electrophysiology is a commonly employed technique; nonetheless, dual whole-cell patch clamp may pose additional challenges. For inexperienced researchers, assistance from an experienced electrophysiologist is recommended, particularly for handling the electrophysiology rig, ensuring proper grounding, and minimizing electrostatic noise. Familiarity with handling two simultaneous electrodes and understanding the amplifier system is prerequisite for successful junctional conductance recordings.

1. Turn on all the necessary equipment and initiate the gravity-fed perfusion and vacuum drain systems. Perfuse the recording chamber with bath solution at a rate of 1–1.5 mL/min.
2. Prepare the octanol solution in the second reservoir to ensure it is ready for perfusion during the experiment.
3. Ensure the bath heating element is activated and maintains a bath temperature of 30 °C.
4. Insert a coverslip with TC620 cells at 80–90% confluency into the recording chamber, ensuring the cells facing upwards (*see* **Note 3**).
5. Utilize the pipette puller to fabricate recording patch clamp pipettes from the borosilicate glass capillary. These pipettes should have a tip resistance ranging from 3–7 mΩ when filled with filtered pipette solution (*see* **Note 4**).
6. Examine the cell culture monolayer in the coverslip piece using the light inverted microscope.
7. Ensure the silver wires from the pipette electrode and ground electrode are appropriately "chlorided" (preferably through electrophoresis using a DC power source within a 0.1 M hydrochloric acid solution) to prevent polarization, offset, and voltage drift during recordings.
8. Insert the ground electrode (*see* **Note 5**) inside a small separate chamber filled with pipette solution, and use a KCl (3 M) agar bridge to connect this ground chamber to the recording chamber. This step is crucial to eliminate junction potential difference during the recordings.

9. Fill two patch pipettes with filtered pipette solution and apply negative pressure (using a vacuum line silicone tube) to remove any air bubble near the tip of the pipette.
10. Carefully insert both pipettes into their respective holders, avoiding scraping the AgCl surface of the electrode. Before immersing the pipettes in the bath solution, maintain a constant slight positive pressure to facilitate a continuous outflow of pipette solution from the pipette, thus preventing contamination.
11. Set the amplifier to voltage clamp mode.
12. Lower the pipettes into the bath solution and apply a small oscillating test pulse (5–10 mV, 30 ms, 50–100 Hz). This should produce a current square response, and the pipette resistance can be measure (test pulse amplitude divided by current response amplitude). This operation is performed automatically performed in the "seal test" tool from pClamp software.
13. Identify a suitable cell pair to patching and secure the microscope stage for the remaining of the procedure.
14. Approach the selected cell pair with both pipettes, using the coarse manipulator initially, and then the micromanipulator as the tips near the cells. Focus the cell membranes.
15. Prior to touching the cell membranes, correct pipette offset with the amplifier (current = 0 pA).
16. Move towards the center of cell-1 (*see* **Note 6**) of the chosen cell pair until the tip makes a small dimple in the membrane. At this time, pipette resistance should increase (lower current response amplitude).
17. Transfer the pressure control system from positive to slight negative pressure to achieve a spontaneous gigaohm seal (<1 GΩ). A successful seal should flatten the current response to the test pulse.
18. Correct the fast capacitance.
19. Repeat **steps 14–16** for cell-2 of the cell pair.
20. Once both cells are in the cell attached mode, brake the membrane by applying negative pressure to the pipettes (*see* **Note 7**).
21. The whole-cell recording mode is reached when the current response to the seal test shows an exponential decay in current right after the voltage step is applied (seal test).
22. If cells are electrically coupled, a current response should be seen in cell-2 when transjunctional voltage is applied (test pulse only applied in cell-1) (*see* **Note 8**).

23. Acquire the recordings with an appropriate software (such as pClamp).
24. The likelihood of finding electrically coupled cell pairs among TC620 cells (80–90% confluency) is approximately 50%.
25. Upon identifying an electrically coupled cell pair, gradually exchange the normal bath solution by 1 mM octanol solution, and confirm that the junctional current is not completely inhibited. If junctional conductance drops less than 25%, then you can consider this experiments as a successful Panx1 cell–cell channel recording. Alternatively, this step can be performed after I–V characterization.
26. Washout the octanol solution by replacing with normal bath solution.
27. Proceed with voltage–clamp mode for I–V characterization. Apply a voltage ramp protocol from −100 to +100 mV (lasting 4–5 s) to cell-1 while recording current responses in cell-2 (keep membrane potential of cell-2 at 0 mV). Repeat this procedure using a voltage ramp protocol from +100 to −100 mV applied in cell-1 (Fig. 1).
28. A dynamic asymmetry in the g_j–V_j relationship should become apparent [10], which is a characteristic of endogenous Panx1 intercellular cell–cell channels expressed in TC620 cells.

4 Notes

1. To prevent cell stress, which can potentially alter the expected results, ensure that the cell growth does not exceed 90% confluence.
2. Dead cells stain blue and should be counted solely for determining cell viability.
3. If the coverslip is too large for the recording chamber, carefully divided it into 2–3 pieces using a precision tweezer. Introduce only one piece into the recording chamber. The remaining pieces can be saved for subsequent experiments.
4. For guidance on creating patch pipettes with the approximate shape and tip resistance, consult the Pipette Cookbook by Sutter instruments (https://www.sutter.com/PDFs/pipette_cookbook.pdf).
5. Ensure that the grounds of both head stage pre-amplifiers are connected to the same ground electrode.
6. It is a common practice to designate the cell on the left side of the cell pair as "cell-1", and the cell at the right side as "cell-2".

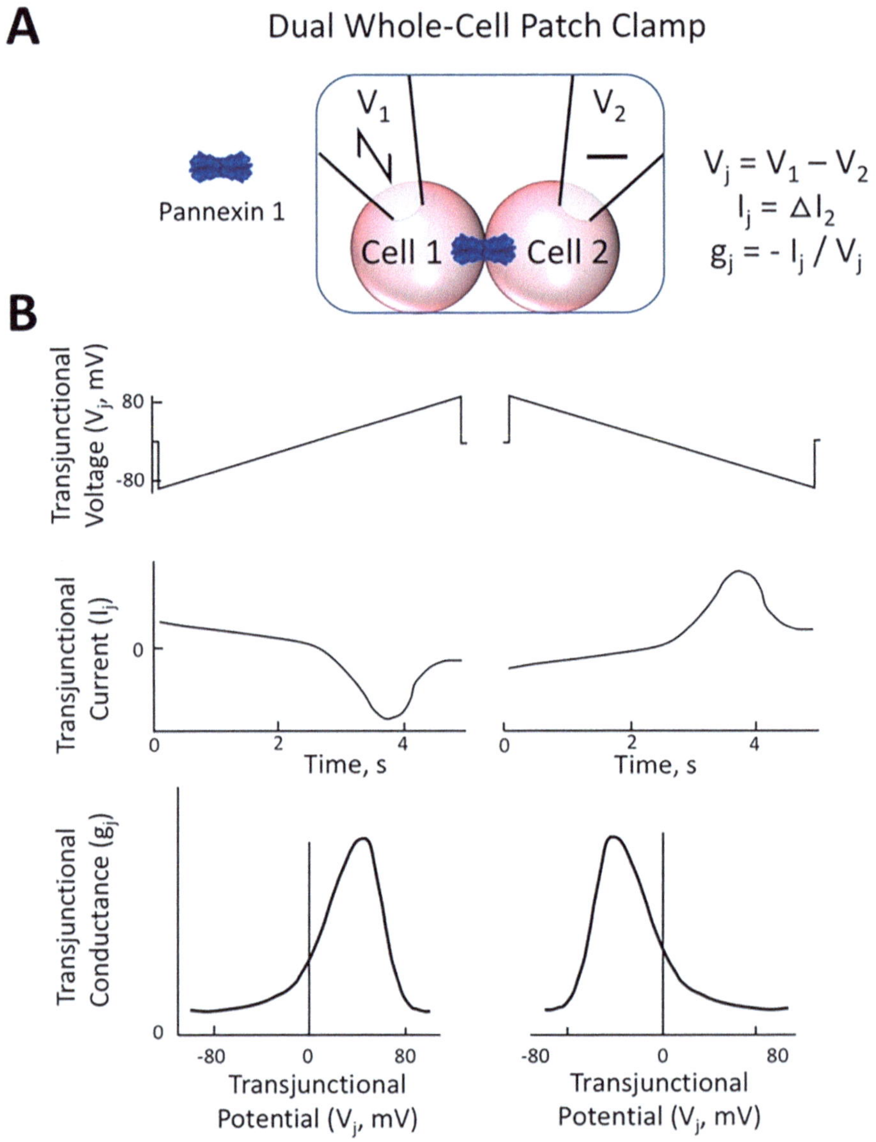

Fig. 1 (**a**) This schematic illustrates a dual whole-cell patch clamp configuration featuring a pannexin 1 intercellular cell–cell channel bridging cell-1 and cell-2. A voltage ramp protocol (V_1) is applied in cell-1, while the voltage of cell-2 (V_2) is maintained constant at 0 mV. The transjunctional voltage (V_j) is produced by stepping the voltage in cell-1 ($\triangle V_1$) and maintaining cell-2 constant, such that $V_j = \triangle V_1$. Transjunctional current (I_j) is measured as the shift in current in the unstepped cell-2, hence $I_j = \triangle I_2$. The transjunctional conductance (g_j) is derived from the current-to-voltage ratio, as $g_j = -I_j/V_j$, where the negative sign denotes that the transjunctional current detected in cell-2 is oriented in the opposite direction to the one measured in cell-1. (**b**) The diagrams display typical transjunctional voltage, current and conductance traces measured in a cell pair expressing pannexin1 cell–cell channels demonstrating a "dynamic asymmetry" in the g_j–V_j relationship (*S-state*). V_j ramps shifting from −100 to +100 mV (left) applied to cell-1 yield an asymmetric g_j–V_j relationship with peak conductance at positive potentials. Conversely, this asymmetry is reversed when V_j ramps from +100 to −100 mV (right) are applied in cell-1, leading to peak conductance at negative potentials

7. The seal membrane can be broken by applying brief pulses of negative pressure using a syringe or by mouth. If it remains intact after a few attempts, the zap function of the amplifier can be employed to deliver a brief electrical discharge through the pipette to brake the seal.
8. The total junctional conductance (g_j) of Panx1 cell–cell channels is relatively low, so increase the gain (up to 100) in cell-B to verify if any current response exists while keeping the membrane potential of cell-B at 0 mV. Sometimes, cell pairs may be electrically coupled solely by a few cell–cell channels. The unitary conductance of the S-state of Panx1 cell–cell channels is ~35 pS [10].

Acknowledgments

This work was partially supported by Fondo Nacional de Desarrollo Científico y Tecnológico grants 3180272 (to N.P.-P.), 1191329 and 1231523 (to J.C.S), as well as Project P09 from the Centro Interdisciplinario de Neurociencias de Valparaíso (to J.C.S).

References

1. Palacios-Prado N, Bukauskas FF (2009) Heterotypic gap junction channels as voltage-sensitive valves for intercellular signaling. Proc Natl Acad Sci USA 106:14855–14860
2. Riol A, Cervera J, Levin M, Mafe S (2021) Cell systems bioelectricity: how different intercellular gap junctions could regionalize a multicellular aggregate. Cancers (Basel) 13:5300
3. Levin M (2021) Bioelectric signaling: reprogrammable circuits underlying embryogenesis, regeneration, and cancer. Cell 184:1971–1989
4. Palacios-Prado N, Bukauskas FF (2012) Modulation of metabolic communication through gap junction channels by transjunctional voltage; synergistic and antagonistic effects of gating and ionophoresis. Biochim Biophys Acta Biomembr 1818:1884–1894
5. Boassa D et al (2007) Pannexin1 channels contain a glycosylation site that targets the hexamer to the plasma membrane. J Biol Chem 282:31733–31743
6. López X et al (2020) Stretch-induced activation of pannexin 1 channels can be prevented by PKA-dependent phosphorylation. Int J Mol Sci 21:9180
7. López X et al (2021) A physiologic rise in cytoplasmic calcium ion signal increases pannexin1 channel activity via a C-terminus phosphorylation by CaMKII. Proc Natl Acad Sci USA 118:e2108967118
8. Bruzzone R, Hormuzdi SG, Barbe MT, Herb A, Monyer H (2003) Pannexins, a family of gap junction proteins expressed in brain. Proc Natl Acad Sci USA 100:13644–13649
9. Sahu G, Sukumaran S, Bera AK (2014) Pannexins form gap junctions with electrophysiological and pharmacological properties distinct from connexins. Sci Rep 4:4955
10. Palacios-Prado N et al (2022) Endogenous pannexin1 channels form functional intercellular cell-cell channels with characteristic voltage-dependent properties. Proc Natl Acad Sci USA 119:e2202104119
11. Sosinsky GE et al (2011) Pannexin channels are not gap junction hemichannels. Channels (Austin) 5:193–197
12. Ambrosi C et al (2010) Pannexin1 and Pannexin2 channels show quaternary similarities to connexons and different oligomerization numbers from each other. J Biol Chem 285:24420–24431
13. Uhlén M et al (2015) Proteomics. Tissue-based map of the human proteome. Science 347:1260419
14. Adamson SE et al (2015) Pannexin 1 is required for full activation of insulin-stimulated glucose uptake in adipocytes. Mol Metab 4:610–618

Chapter 12

Generation of Connexin-Expressing Stable Cell Pools

Abraham Tettey-Matey, Chiara Di Pietro, Viola Donati, Fabio Mammano, and Daniela Marazziti

Abstract

Stable cell pools have the advantage of providing a definite, consistent, and reproducible transmission of a transgene of interest, compared to transient expression from a plasmid transfection. Stably expressing a transgene of interest in cells under induction is a powerful way to (switch on and) study a gene function in both in vitro and in vivo assays. Taking advantage of the ability of lentivirus (LV) to promote transgene delivery, and genomic integration and expression in both dividing and nondividing cells, a doxycycline-inducible transfer vector expressing a bicistronic transgene was developed to study the function of connexins in HeLa DH cells. Here, delving on connexin 32 (Cx32), we report how to use the backbone of this vector as a tool to generate stable pools to study the function of a gene of interest (GOI), especially with assays involving Ca^{2+} imaging, employing the GCaMP6s indicator. We describe a step-by-step protocol to produce the LV particle by transient transfection and the direct use of the harvested LV stock to generate stable cell pools. We further present step-by-step immunolabeling protocols to characterize the transgene protein expression by confocal microscopy using an antibody that targets an extracellular domain epitope of Cx32 in living cells, and in fixed permeabilized cells using high affinity anti-Cx32 antibodies. Using common molecular biology laboratory techniques, this protocol can be adapted to generate stable pools expressing any transgene of interest, for both in vitro and in vivo functional assays, including molecular, immune, and optical assays.

Key words Lentiviral transduction, Hemichannels, Gap junction channels, Immunofluorescence, Confocal microscopy

1 Introduction

Parental HeLa DH cells are a human cervix carcinoma cell line (European Collection of Authenticated Cell Cultures, ECACC, Cat. No. 96112022) that have been widely used for transfection assays using wild type or mutant connexins and subsequent analysis of gap junction formation, permeability of intercellular gap junction channels and hemichannels, and sensitivity to pharmacological modulator (see, e.g., [1–4]). HeLa DH cells exhibit stable membrane potentials, high stability under double whole-cell patch-

Fabio Mammano and Mauricio Retamal (eds.), *Connexin Hemichannels: Methods and Protocols*, Methods in Molecular Biology, vol. 2801, https://doi.org/10.1007/978-1-0716-3842-2_12,

clamp measurements, and low levels of Cx45 expression [5]. This chapter provides detailed information to generate stable pools of HeLa DH cells expressing the connexin of interest, delving into Cx32 (for a list of human and mouse connexins, *see* Table 1). Mutations in the *GJB1* gene, which encodes Cx32, are the leading cause of X-linked Charcot–Marie–Tooth disease (CMT1X), the second most prevalent (~1:50,000) connexin-related genetic disease, for which there is no cure [6].

The present protocol is based on the lentiviral transfer-vector plasmid pCW57-MCS-IRES-GCaMP6s used by Nardin et al. [7]. This plasmid contains the following cis-acting genetic elements derived from the HIV-1 [8–11]: two long terminal repeats (LTRs) at the 5′ and 3′ untranslated regions (UTRs) for host genomic integration; a packaging signal (Ψ) to promote the production and packaging of only the transfer vector as the virus genome; a Rev. response element (RRE) for rev protein-dependent export of un-spliced transfer vector (RNA genome) from the host nucleus to the cytoplasm [9, 11], and central poly-purine tract/central termination sequence (cPPT/CTS) that promotes reverse transcription and importation of the viral vector into the host nucleus [12]. Here, our GOI, the human Cx32 (*GJB1*, NCBI CCDS ID CCDS14408.1) was inserted between EcoRI and MluI sites of the pCW57-MSC-IRES-GCaMP6s backbone to form the lentiviral vector plasmid used for these experiments (Fig. 1).

The vector also contains non–HIV-1 genetic elements such as a tetracycline-inducible promoter (tight TRE-promoter), a multiple cloning site (MSC) located upstream of Encephalomyocarditis virus (EMCV) internal ribosome entry site (IRES, gray) and the coding sequence of GCaMP6s (green) [13]. The human phosphoglycerate kinase (hPGK) promoter is located to the downstream side. It constitutively drives the expression of puromycin N-acetyl transferase (PuroR, red) linked by a *Thosea asigna* virus capsid 2A (T2A, dark) peptide sequence to that of the advanced tetracycline-controlled transactivator (rtTA-Advance, violet) proteins [14]. Failure of a peptide bond formation between amino acids glycine and proline in the T2A peptide, the bicistronic puromycin-T2A-rtTA-Advance sequence produced separate proteins. The PuroR confers resistance to the antibiotic puromycin added to the cell culture medium, allowing the selection and generation of stable cell pools [15]. The rtTA-Advance protein binds upon the activation by doxycycline (dox) to the tetracycline (Tet) responsive element (TRE) in the tight TRE promoter to switch on the gene of interest (GOI) expression from the MCS.

The bicistron Cx32-IRES-GCaMP6s is expressed under the tight TRE-promoter, upon induction with dox addition to the culture medium. The EMCV IRES sequence promotes cap-independent

Table 1
List of human and mouse connexin attributes

#	*Homo sapiens* Gene Symbol	Protein Name	OMIM ID	NCBI CCDS Database ID	NCBI Reference Sequence ID	*Mus musculus* Gene Symbol	Protein Name	MGI ID	NCBI CCDS Database ID	NCBI Reference Sequence ID
1	*GJA1*	CX43	121014	CCDS5123.1	NM_000165.5	*Gja1*	Cx43	95713	CCDS23851.1	NM_010288.3
2	*GJA3*	CX46	121015	CCDS9289.1	NM_021954.4	*Gja3*	Cx46	95714	CCDS27152.1	NM_016975.3
3	*GJA4*	CX37	121012	CCDS30669.1	NM_002060.3	*Gja4*	Cx37	95715	CCDS18667.1	NM_008120.3
4	*GJA5*	CX40	121013	CCDS929.1	NM_181703.4	*Gja5*	Cx40	95716	CCDS17652.1	NM_008121.3
5	–	–	–	–		*Gja6*	Cx33	95717	CCDS30506.1	NM_001001496.2
6	*GJA8*	CX50	600897	CCDS30834.1	NM_005267.5	*Gja8*	Cx50	99953	CCDS17651.1	NM_008123.3
7	*GJA9*	CX59	611923	CCDS432.1	NM_030772.5	–	–	–	–	
8	*GJA10*	CX62	611924	CCDS5025.1	NM_032602.2	*Gja10*	Cx57	1339969	CCDS18015.1	NM_010289.2
9	*GJB1*	CX32	304040	CCDS14408.1	NM_001097642.3	*Gjb1*	Cx32	95719	CCDS30314.1	NM_008124.3
10	*GJB2*	CX26	121011	CCDS9290.1	NM_004004.6	*Gjb2*	Cx26	95720	CCDS27153.1	NM_008125.3
11	*GJB3*	CX31	603324	CCDS384.1	NM_024009.3	*Gjb3*	Cx31	95721	CCDS18668.1	NM_008126.2
12	*GJB4*	CX30.3	605425	CCDS383.1	NM_153212.3	*Gjb4*	Cx30.3	95722	CCDS18669.1	NM_008127.4
13	*GJB5*	CX31.1	604493	CCDS382.1	NM_005268.4	*Gjb5*	Cx31.1	95723	CCDS18670.1	NM_010291.3
14	*GJB6*	CX30	604418	CCDS9291.1	NM_006783.5	*Gjb6*	Cx30	107588	CCDS27154.1	NM_001271663.1
15	*GJB7*	CX25	611921	CCDS5008.1	NM_198568.3	–	–	–	–	
16	*GJC1*	CX45	608655	CCDS11487.1	NM_005497.4	*Gjc1*	Cx45	95718	CCDS25504.1	NM_001359040.1
17	*GJC2*	CX47	608803	CCDS1569.1	NM_020435.4	*Gjc2*	Cx47	2153060	CCDS24760.1	NM_080454.4
18	*GJC3*	CX29/30.2/31.3	611925	CCDS34697.1	NM_181538.3	*Gjc3*	Cx29	2153041	CCDS19782.1	NM_080450.5

(continued)

Table 1
(continued)

	Homo sapiens					*Mus musculus*				
#	**Gene Symbol**	**Protein Name**	**OMIM ID**	**NCBI CCDS Database ID**	**NCBI Reference Sequence ID**	**Gene Symbol**	**Protein Name**	**MGI ID**	**NCBI CCDS Database ID**	**NCBI Reference Sequence ID**
19	*GJD2*	CX36	607058	CCDS10040.1	NM_020660.3	*Gjd2*	Cx36	1334209	CCDS16563.1	NM_010290.2
20	*GJD3*	CX31.9	607425	CCDS58547.1	NM_152219.4	*Gjd3*	Cx30.2	2384150	CCDS25369.1	NM_178596.3
21	*GJD4*	CX40.1	611922	CCDS7191.1	NM_153368.3	*Gjd4*	Cx39	2444990	CCDS29049.1	NM_153086.5
22	*GJE1*	CX23	611925	CCDS87449.1	NM_001358410.2	*Gje1*	Cx23	1923993	CCDS48505.1	NM_029722.2

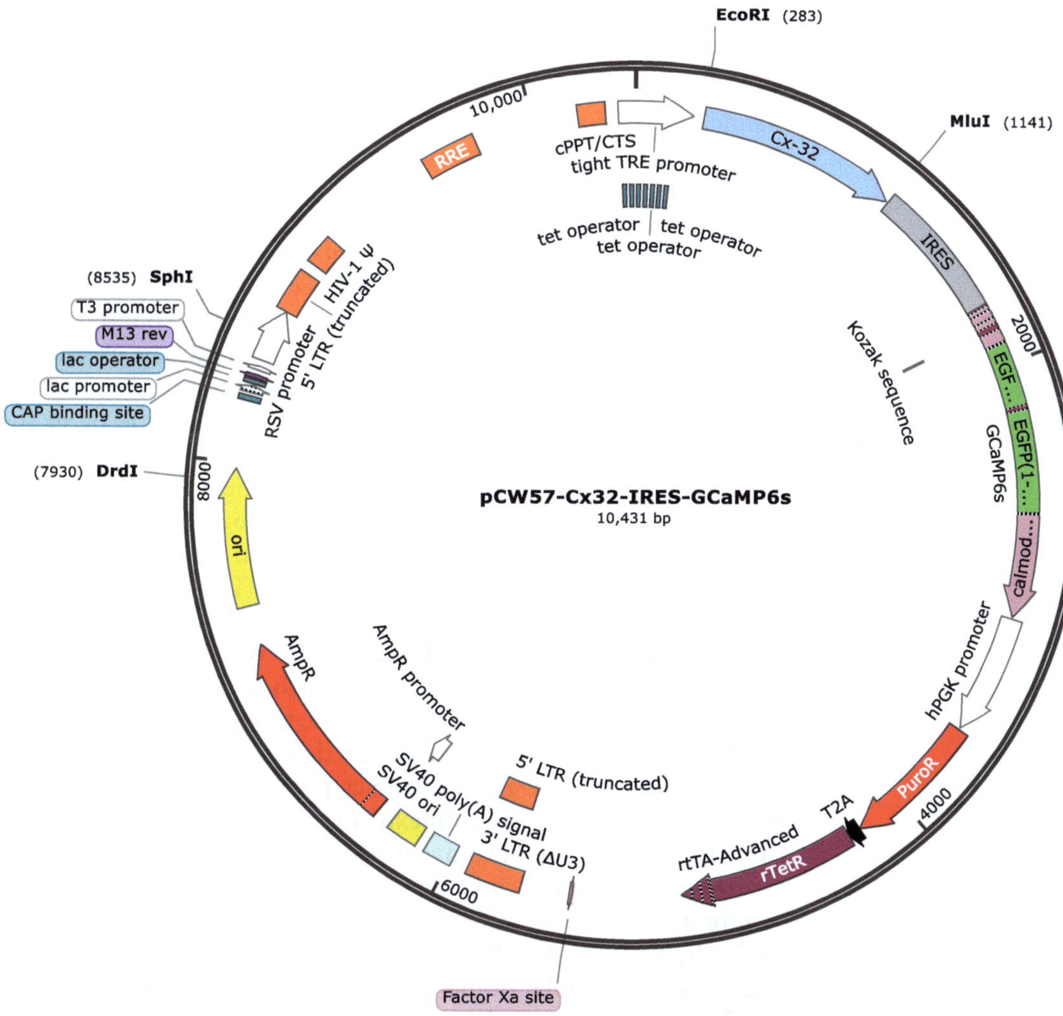

Fig. 1 Lentiviral transfer plasmid for the bicistronic expression of Cx32 and cytosolic GCaMP6s. Schematic map of the lentiviral vector transfer plasmid expressing the genes of interest (GOIs) under the tetracycline-inducible promoter (tight TRE-promoter). For controlled selection of stable pools, the pCW57.1 vector (a mammalian expression, lentiviral, Tet-on, destination vector) was modified by inserting a puromycin selection marker under the constitutive, human phosphoglycerate kinase (hPGK) promoter. The coding sequence (CDS) of human Cx32 linked to an IRES and GCaMP6s sequences was gene synthesized and subcloned using the EcoRI and MluI sites of the vector, allowing the independent but simultaneous expression of Cx32 and cytosolic GCaMP6s proteins from the transfer vector

translation of the GCaMP6s [13], allowing for independent production of the Cx32 and GCaMP6s, as separate proteins. GCaMP6s can be used as fluorescent marker, signaling response to dox of transduced cells, and/or to conduct quantitative assays of Ca^{2+} uptake through connexin hemichannels [7]. Thus, the pCW57-Cx32-IRES-GCaMP6s lentiviral transfer-vector plasmid with the Tet-on system is a tool to controllably induce the expression of Cx32 and GCaMP6s proteins for functional analyses in vitro as well as in vivo. The use of an inducible promoter is particularly helpful when dealing with hemichannels, as uncontrolled expression of leaky or hyperactive hemichannels causes toxicity and rapidly leads to cell death (see, e.g., [16–23]).

In the absence of dox addition to the culture, there is some leakiness from the tight TRE-promoter, hence a background Tet-on activation, leading to basal detection levels of the Cx32 and the GCaMP6s expression. To reduce this background leakage, a tetracycline-free FBS (Tet System approved) should be used in the culture medium, especially during viral production to increase the titer, and when studying a transgene mutation that is toxic to the generated stable cell line.

Like all plasmids, important genetic sequences such as SV40 and high-copy-number ColE1/pMB1/pBR322/pUC origin of replication and β-lactamase are also present in the pCW57-Cx32-IRES-GCaMP6s plasmid backbone [24–26]. The β-lactamase confers resistance to ampicillin for the selection of transformed bacteria. The ColE1/pMB1/pBR322/pUC origin of replication sequence confers a high copy number of the plasmid production.

In this chapter, we explain how to use the pCW57-Cx32-IRES-GCaMP6s plasmid to produce LV particles (Fig. 2).

The latter can be used to infect parental HeLa DH cells to generate stable pools of connexin expressing cells (HeLa-Cx32-GCaMP6s) following puromycin selection (Fig. 3).

To confirm that hemichannels were expressed on the plasma membrane of HeLa-Cx32-GCaMP6s cells following dox induction for 72 h, we used the abEC1.1 monoclonal antibody (mAb) that targets an extracellular domain epitope shared by Cx26, Cx30, and Cx32 HCs with nanomolar sensitivity and has been extensively characterized both in vitro and in vivo [7, 27–32] (Fig. 4).

In HeLa-Cx32-GCaMP6s cells induced with dox for 72 h, overexpressed connexins were also detected by conventional immunofluorescence staining with a commercial antibody that recognizes an intracellular epitope. Only in this case, immunoreactivity produced fluorescent puncta at points of contact between adjacent cells, corresponding to intercellular gap junction channels (IGJCs, Fig. 5).

By replacing the coding sequence of Cx32 in the pCW57-Cx32-IRES-GCaMP6s plasmid with that of any other connexin (wild type, *see* Table 1, or mutant), the instructions provided herein shall enable its controlled expression in HeLa DH cells, or any other cell type amenable to LV transduction, and the generation of a stable cell pool. Cell sorting using the GCaMP6s reporter can also be used to further select cells with the highest expression levels.

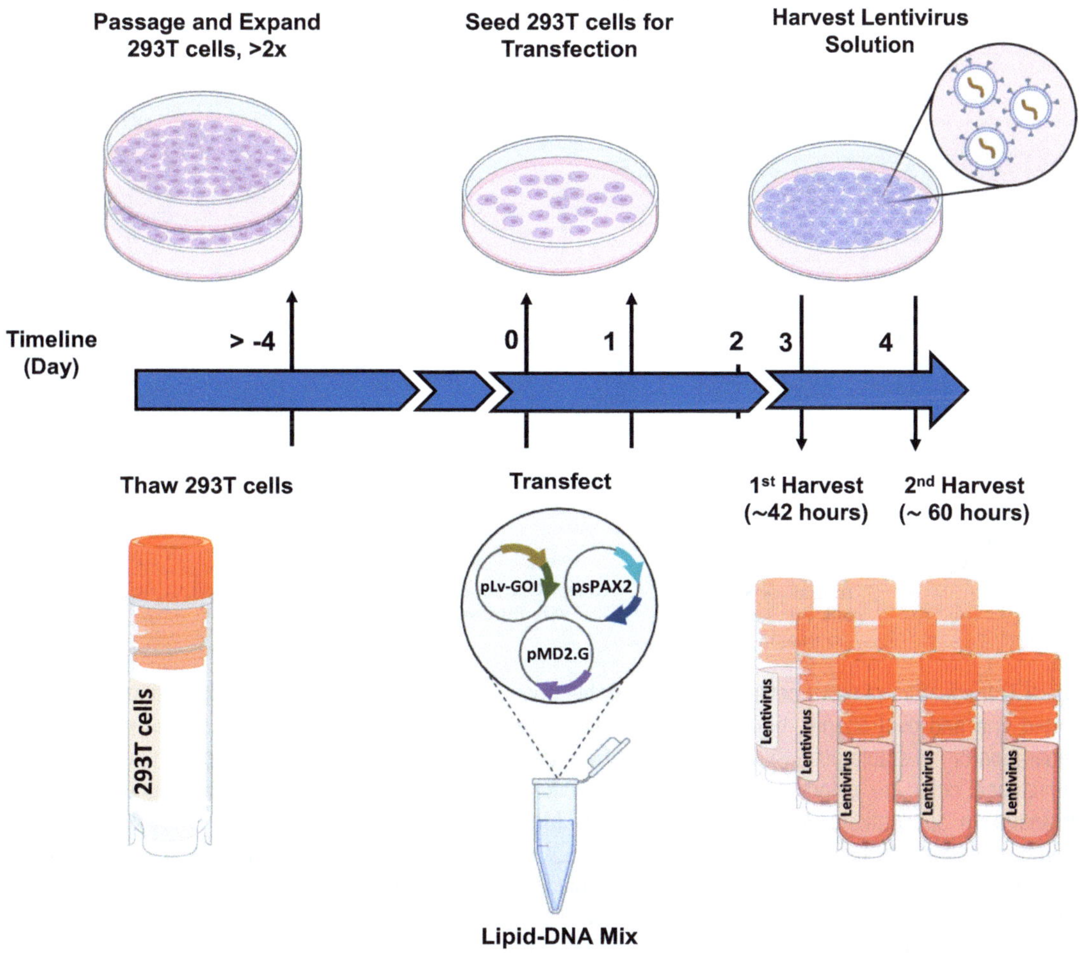

Fig. 2 Overview of LV production. Timeline (in blue) for the LV production, as shown in the scheme diagram. Thaw and expanding 293T cells takes more than 4 days. Cells should be passaged at least twice (2×) before using them for virus production. On day 0, seed 293T cells to be >85% confluent 24 h later, ready for the transfection. On day 1, transfect cells using a cocktail of the LV production plasmid DNA mix: the lentiviral transfer-vector plasmid (pLV-GOI) containing the genes of interest (GOIs), the packaging (psPAX2), and envelope (pMD2.G) plasmids. If lipofectamine 3000 was used for transfecting the plasmids into the cells, replace the medium 5–6 h later. On day 2, 18 h post-transfection, refresh the medium in preparation for virus harvesting. On day 3, 42 h post-transfection, collect the virus supernatant (first harvest) from the culture, refresh the medium and continue the incubation. Make aliquots, 1 mL of the filtered supernatant, and use them directly for transduction, or freeze them for long-term storage. On day 4, 60 h post-transfection, perform the second harvest of viral particles and discard the cells

2 Materials

Handle materials, reagents, and working areas following aseptic techniques and standard practices. For the LV production and generation of stable pools, work in a sterile biosafety cabinet level 2 (*see* **Note 1**).

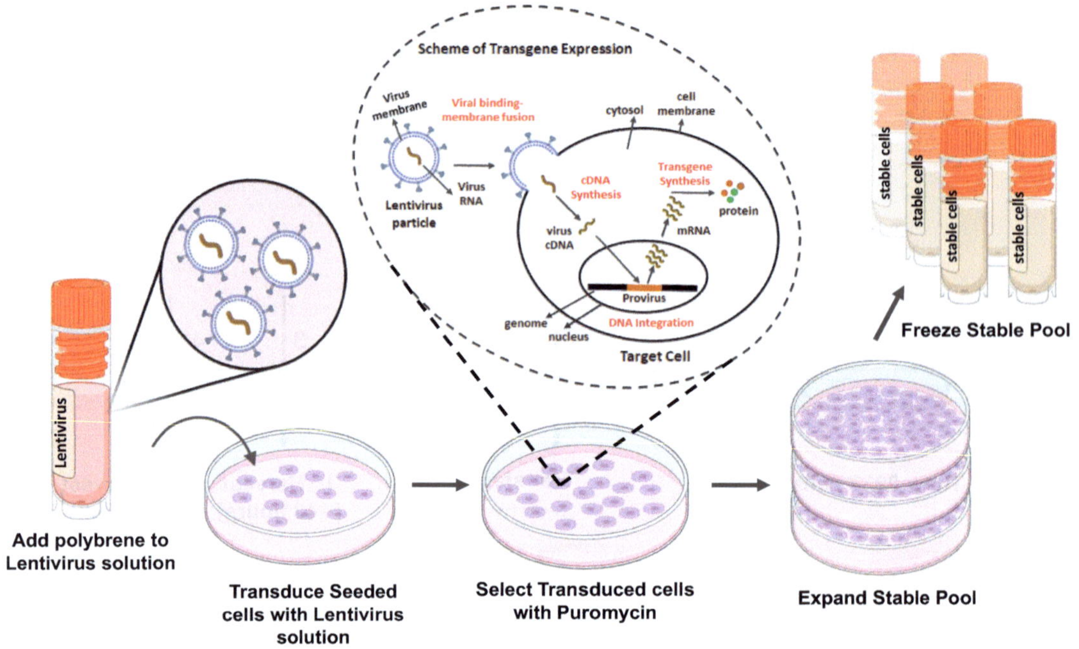

Fig. 3 Transduction of naïve cells to generate stable pools. Scheme of the transgene expression (shown as a dotted line and oval shape), highlighting the key pathways (in red), leading to positive cell selection and generation of the stable pool. Seeded naïve cells are transduced with the LV stock mixed with polybrene. The transduced cells are selected with puromycin to form a stable pool. The stable pool is expanded for downstream applications and storage. Upon cell infection with the LV solution, the virus and host cell membranes interact, leading to the release of the LV transfer vector (depicted in brown), as the virus RNA genome into the naïve cell. This viral RNA is copied into cDNA and then integrated into the cell genome (depicted in black). The puromycin selection marker is expressed from the integrated transfer vector genome to select positive cells resistant to puromycin added to the cell culture. Other genes in the transfer vector are expressed and can be detected by RNA and protein expression analysis. DNA (deoxyribonucleic acid), cDNA (complementary DNA), RNA (ribonucleic acid), mRNA (messenger RNA)

2.1 Cell Culture, Cell Transfection, LV Production, Stable Pools Generation and Cryopreservation

1. Complete growth medium for targeted HeLa cells culturing: in a sterile biosafety cabinet, supplement Dulbecco's Modified Eagle Medium DMEM (high glucose 4.5 g/L, sodium pyruvate 110 mg/L) with 10% heat-inactivated tetracycline-free FBS (TSA-FBS), 2 mM L-glutamine, and 1% penicillin/streptomycin to a total volume of 500 mL. After mixing by inverting the closed bottle 4–6 times, the complete DMEM is warmed to 37 °C before using (*see* **Note 2**).
2. Complete growth medium for Lenti-X 293T viral production: in a sterile biosafety cabinet, supplement DMEM (high glucose 4.5 g/L, sodium pyruvate 110 mg/L, GlutaMAX supplement) with 10% heat-inactivated tetracycline-free FBS (TSA-FBS) (*see* **Note 3**), 6 mM L-glutamine, 1 mM sodium pyruvate, 25 mM hepes, and 1% penicillin/streptomycin to a total volume of 500 mL. Mix by inverting the closed bottle 4–6 times and warm to bring the complete DMEM to 37 °C before using (*see* **Note 2**).

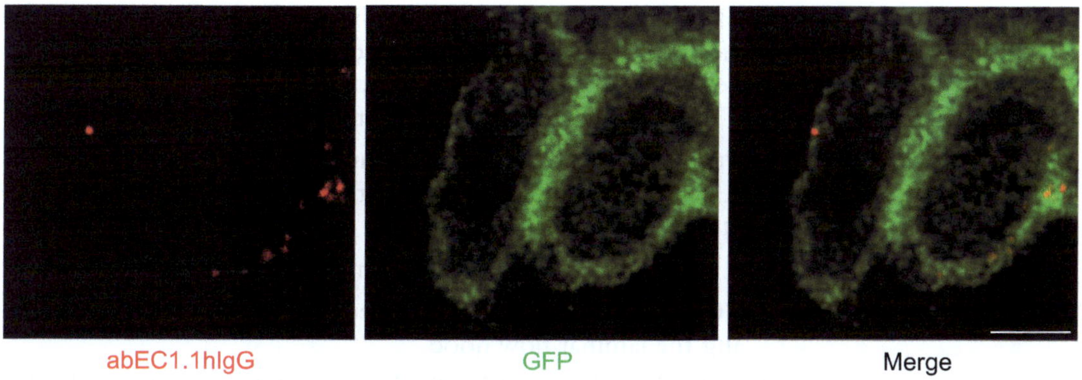

Fig. 4 In vivo immunolabeling of HeLa cells to detect Cx32 hemichannels
Representative confocal images of abEC1.1hIgG (red) and GFP (green) in HeLa cells expressing Cx32. The abEC1.1hIgG antibody (5 nM) that binds the extracellular domain of Cx32 HCs was applied to live cells and counterstained with a red-fluorescent secondary antibody selective for the human Fc domain of the IgG. HeLa cells expressing Cx32 were identified by anti-GFP staining; scale bar: 25 μm

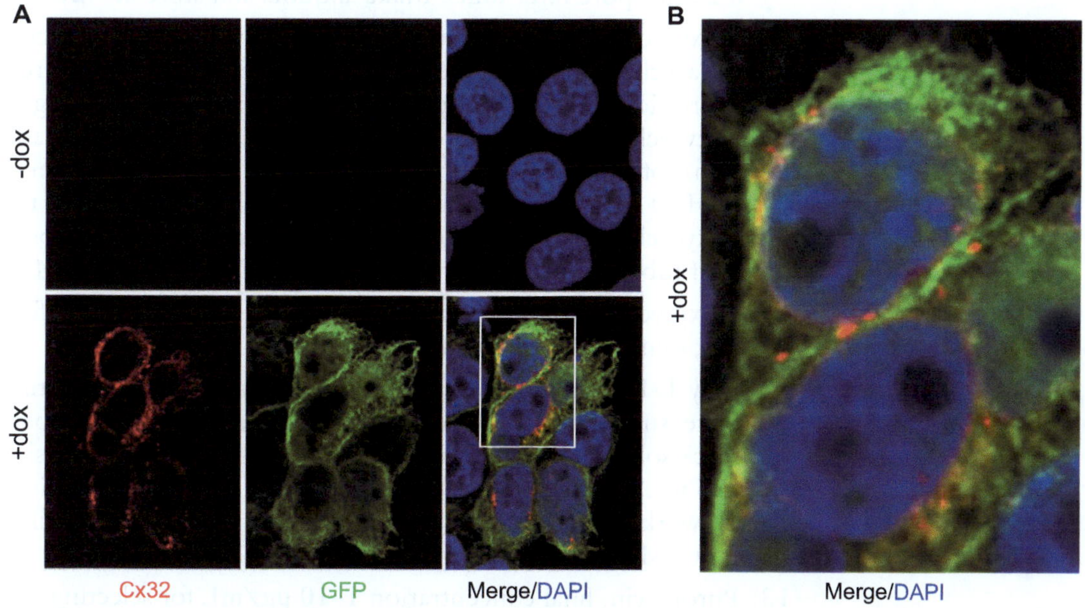

Fig. 5 Immunolabeling of HeLa cells to detect whole Cx32 expression
(**a**) Representative confocal images of fixed and permeabilized HeLa cells labeled with anti-Cx32 (red) and anti-GFP (green) antibodies; scale bar: 10 μm. (**b**) Higher magnification view of boxed area (white rectangle) showing fluorescent puncta at points of contacts between adjacent cells; scale bar: 5 μm

3. Stock solutions of trypsin-EDTA (0.25% trypsin, 2.2 mM EDTA tetrasodium) for detaching cells during seeding or passaging are stored in 50 mL aliquots at −20 °C. The working aliquot is thawed and stored at 4 °C.

4. Sterile phosphate-buffered saline (1× PBS, pH 7.4, without calcium) for washing away dead cells and debris from plated cell surface monolayer before passaging or seeding the cells. The working aliquot is stored at 4 °C.
5. Sterile OptiMEM reduced serum media, for preparing transfection DNA/lipid mix, cell transfection, and diluting LV stocks for transduction.
6. Lipofectamine 3000 transfection reagent.
7. 70% ethanol to clean and disinfect materials and surfaces entering the laminar flow hood.
8. Polyethersulfone (PES) 0.45 μm-pore filter unit for clearing LV stocks free of 293T cells and debris.
9. Polybrene (Hexadimethrine bromide), final concentration 8 μg/mL, for LV transduction. Prepare a solution of 4 mg/mL, by dissolving 4 mg in 1 mL of sterile ddH_2O under the hood, to obtain a 1000× stock solution. Filter sterilize it with 0.22 μm-pore filter units. Make aliquots and store at −20 °C.
10. Doxycycline (dox) for activating the transfer vector tetracycline-inducible promoter for targeted gene of interest expression. Prepare 1 mg/mL stock solution: weigh 10 mg of doxycycline monohydrate in a 1.5 mL Eppendorf tube, add 1 mL of DMSO and vortex the mixture. Add 9 mL of sterile ddH_2O to a 15 mL conical centrifuge tube and mix with the doxycycline solution. Aliquot the solution in 1.5 mL Eppendorf tubes and store them at −20 °C for 1 year (*see* **Note 4**).
11. Doxycycline growth medium (4 μg/mL): immediately before use, dilute 1:250 in growth medium.
12. Poly-L-Lysine final concentration, 0.1 mg/mL for coating plate surfaces to enhance seeded cells attachment. Prepare stock solution of 1 mg/mL, by dissolving 5 mg Poly-L-Lysine in 5 mL of sterile ddH_2O under the hood, such that there is 1:10 working dilution. Filter sterilize it with 0.22 μm-pore filter units. Make 1 mL aliquots and store at −20 °C.
13. Puromycin, final concentration 1–10 μg/mL for selecting stable cell lines. If in powder form, prepare stock of 10 mg/mL by dissolving 10 mg in 1 mL of sterile ddH_2O. Prepare prewet 0.22 μm-pore filter units, by passing 1 mL of sterile ddH_2O through it. Then, pass the puromycin stock solution though to sterilize it. Make 100 μL aliquots and store at −20 °C.
14. Cell culture grade dimethyl sulfoxide (DMSO), for preparing dox stock solution and cell line freezing medium (10% DMSO in complete medium).
15. HEK 293T-derived cell line optimized for LV production, Lenti-X 293T cell line.

16. Transfer LV vector plasmid pCW57-Cx32-IRES-GCaMP6s (for details, *see* Fig. 1).
17. Third-generation LV envelope and packaging vector plasmids, pMD2.G and psPAX2.
18. Sterile 100, 60, and 35 mm tissue-culture dishes.
19. 5¾″ borosilicate glass Pasteur pipette (autoclaved).
20. Sterile cryovial tubes for cryopreservation of LVes and cells.
21. Biological Safety Cabinet, preferably a Class II-type, BSC/laminar flow hood (with a HEPA microfilter).
22. Protective laboratory equipment, such as a lab coat and appropriate gloves.

2.2 HeLa Cells Immunolabeling

1. Sterile 35 mm tissue-culture dishes and 24 mm well plates.
2. Borosilicate glass microscope coverslips (12 mm diameter, autoclaved).
3. Rocking platform.
4. Dark plastic box with lid.
5. Whatman filter paper.
6. Eppendorf tubes.
7. Eppendorf benchtop centrifuge.
8. Forceps.
9. Glass microscope slides.
10. Microscope slide folder.
11. ProLong™ Gold antifade mountant.
12. Sterile phosphate-buffered saline (1× PBS, pH 7.4).
13. abEC1.1hIgG diluting solution: 138 mM NaCl, 5 mM KCl, 0.4 mM $NaHPO_4$, 10 mM HEPES, 6 mM D-glucose, 60 μM $CaCl_2$, pH 7.2/7.3, 320 mOsm.
14. abEC1.1hIgG solution: dilute to 5 nM abEC1.1hIgG in diluting solution.
15. Fixing solution: 4% paraformaldehyde in 1× PBS.
16. Permeabilization solution: 0.1% Triton X-100 in 1× PBS.
17. Blocking solution: 0.5% BSA, 0.1% Triton X-100 in 1× PBS. Add 1 mL 1× PBS, to a 15 mL conical centrifuge tube. Add 0.2 g of bovine serum albumin (BSA), add 50 μL of 20% Triton X-100, and top up to 10 mL with 1× PBS. Vortex mixture (*see* **Note 5**). Store at 4 °C.
18. Primary antibody dilution: dilute the appropriate primary antibody in blocking solution.
19. Secondary antibody dilution: dilute the appropriate secondary antibody 1:800 in 1× PBS (*see* **Note 6**).

20. DAPI stock solution: dissolve 5 mg/mL of 4′,6-diamidino-2-phenylindole (DAPI) in 1× PBS to obtain a 1000× stock solution. Prepare 25 μL aliquots and store −20 °C (*see* **Note 7**).
21. DAPI working solution: thaw one aliquot of DAPI stock solution and dilute 1:1000 in 1× PBS.

2.3 Leica TCS SP5 Inverted Confocal Microscope

The Leica TCS SP5 inverted microscope is equipped with three photomultiplier tubes (PMT) detectors that allow for three emission spectrums from live or fixed cells to be acquired simultaneously or sequentially and separated without overlapping signal bleed through and image contamination. The addition of a transmitted light detector increases the breadth of images that can be acquired. The system we use is equipped with four class 3B lasers:

- Diode 405 nm (Maximum luminous power at laser output <60 mW, Maximum luminous power in focal plane <6 mW).
- Argon 488 nm (Maximum luminous power at laser output <200 mW, Maximum luminous power in focal plane <30 mW).
- HeNe 543 nm (Maximum luminous power at laser output <1.5 mW, Maximum luminous power in focal plane <0.5 mW).
- HeNe 633 nm (Maximum luminous power at laser output <15 mW, Maximum luminous power in focal plane <4 mW).

These lasers allow for the fluorescence emitted from most blue, green, and red fluorochromes and a wide variety of fluorescent proteins.

The inverted microscope is fitted with a suite of objective lenses that includes:

- 20× dry objective (HC PL FLUOTAR, UV optimized, NA 0.5, Leica)
- 40× oil immersion objective (HC PL Apo, UV optimized, NA 1.25, Leica)
- 63× oil immersion objective (HC PL Apo, UV optimized, NA 1.4, Leica).

Images are captured on Leica Application Suite Advanced Fluorescence 2.7.3.9723 (LAS AF) software by Leica Microsystems and archived on a connected server (*see* **Note 8**).

3 Methods

3.1 Plating and Culturing Lenti-X 293T Cells for Transfection

1. Passage and grow Lenti-X 293T cells (*see* **Note 9**), when 80% confluent in 10 mL of complete DMEM cell culture medium in a 100 mm sterile plate. Maintain the cells in a 5% CO_2 humidified incubator at 37 °C.

2. For general maintenance, split the cells 1/10 (1.0×10^6 cells) in a 100 mm sterile plate twice a week.
3. For transient transfection to produce LV, split about 4×10^6 cells to achieve a cell density of 80–90% confluent at the time of transfection, using Lipofectamine 3000 transfection reagent.
4. One day (day 0, Fig. 1) before transfection, precoat the 100 mm sterile plate with 5 mL of 0.1 mg/mL poly-L-lysine for 30 min at 37 °C (*see* **Note 10**), aspirate and wash excess poly-L-lysine twice with 2.5 mL of 1× PBS at room temperature and dry the plates in the hood under UV light.
5. Meanwhile, prewarm 1× PBS, trypsin-EDTA, and complete DMEM in the water bath for about 15 min at 37 °C for plating Lenti-X 293T (*see* **Note 2**). Pick the ~80% confluent plate from the incubator and carefully aspirate the cell culture growth medium, and gently rinse the cell monolayer with 5 mL of 1× PBS to remove cell debris and dead cells. Trypsinize the cells for 3 min with 1 mL of 0.05% trypsin-EDTA at the hood temperature. Rock the plate to dislodge cells, resuspend cells, and inactivate trypsin with 4 mL of complete DMEM for Lenti-X 293T cells.
6. Transfer the suspension into a clean 15 mL conical centrifuge tube, and pellet the cells for 3 min at 250 × *g*. Carefully aspirate and discard the supernatant. With p1000 filtered tips, add 1 mL of fresh medium, and resuspend the cells by pipetting up and down 10–20 times to break the cell pellets into single cells in suspension. Top it up to 10 mL of complete DMEM.
7. Determine the cell concentration using a hemocytometer or any appropriate cell counting tool, and plate about 4×10^6 cells or an equivalent to reach 80–90% confluent at the time of transfection. Adjust the culture medium to 10 mL, and gently rock the plate back and forth to mix and distribute the cells. Then, tilt the plate to one side and pipette up and down 3–4 times to uniformly distribute the cells over the growth surface area (*see* **Note 11**). Gently handle and incubate the plated cells at 37 °C in a 5% CO_2 humidified incubator.

3.2 Transfecting Lenti-X 293T Cells (for LV Production)

1. On day 1 at the time of transfection, prewarm OptiMEM I (Reduced Serum Medium) at 37 °C. In the meantime, thaw and bring aliquots of the 3 Lentiviral plasmid DNA to room temperature, prior to use.
2. Following aseptic techniques, prepare the co-transfection mixes with the transfecting reagent Lipofectamine 3000 or a similar one, using two sterile 1.5 mL Eppendorf tubes (as indicated below).

Tube 1: OptiMEM/DNA/p3000 mix, to a total of 600 μL.

Transfer plasmid, pCW57-Cx32-IRES-GCaMP6s: 10 μg.

psPAX2 (Addgene plasmid #12260): 9.5 μg.

pMD2.G (Addgene plasmid #12259): 4.5 μg.

P3000 reagent: 55 μL.

Vortex briefly, for 3 s to mix, and preincubate Tube 1 mix for 5 min in the hood.In the meantime, prepare:

Tube 2: OptiMEM/Lipo3000 mix, to a total of 600 μL.

Lipofectamine 3000 transfection reagent: 65 μL

3. Mix 600 μL of Tube 1 to 600 μL of Tube 2 (*see* **Note 12**), vortex briefly for 6 s to mix, and spin down the mixture briefly. Incubate the final mix for 35 min in the dark hood at room temperature.
4. Prior to transfecting the DNA-lipid (Lipofectamine 3000) complex, carefully aspirate and replace the 10 mL culture medium with 4 mL of prewarm OptiMEM at 37 °C.
5. Carefully add the 1200 μL mixture dropwise, spreading it across the surface area without disrupting the cell monolayer. Gently agitate the plate to evenly distribute the DNA-lipofectamine complex for increased transfection efficiency. Incubate the transfected cell culture at 37 °C in a 5% CO_2 humidified incubator.
6. Replace the transfected culture containing the DNA-lipid complex at 5 h post-transfection (*see* **Note 13**). Carefully aspirate the transfection culture in the hood and discard it into a container with 20% bleach. Replace it with a prewarmed complete DMEM for Lenti-X 293T culturing and LV production (*see* **Note 14**).
7. On the next day (Day 2), carefully replace the 10 mL medium within 18 h post-transfection with a 37 °C prewarmed complete DMEM for Lenti-X 293T cells. Discard the waste aseptically (*see* **Note 14**). If the transfer lentiviral plasmid contains a reporter gene such as GCaMP6s, monitor its expression with fluorescence microscopy (*see* **Note 15**).
8. On day 3 (42 h post-transfection, Fig. 2), aseptically harvest the 10 mL medium containing the LV into a sterile 15 mL tube and replace it with a fresh 10 mL prewarmed complete medium. Return the culture to the incubator and continue the incubation. Centrifuge the collected medium at 500 × *g* for 5 min to pellet cell debris, collect and filter supernatant into a fresh 15 mL tube to completely remove all cell debris, using a 0.45 μm PES filter (*see* **Note 16**).

9. Aliquot 1 mL of the filtered supernatant into a sterile cryovial for immediate use in virus transduction test or storage at 4 °C temporarily or −80 °C (*see* **Note 17**).
10. On day 4 (60 h post-transfection), perform the second virus harvesting, as before (*see* **step 8**), but aseptically discard the producing Lenti-X 293T cells. For a highly concentrated virus, pull both harvest 1 and 2 supernatants together, 20 mL, and concentrate the virus. Resuspend the virus in a small volume of 1× PBS and store aliquots of 10–100 μL in sterile 0.5 mL tubes at −80 °C (Fig. 2).

3.3 LV Stock Viability Test by Puromycin Selection in Lenti-X 293T Cells

1. One day before viral infection, seed three wells in a 6 multiwell plate. Two of the wells will be used to test viral transduction efficiency using 0.5 and 1 mL of the collected LV stock and one well for negative control, no infection. Plate 1.5×10^5 Lenti-X 293T cells per well in a complete DMEM for Lenti-X 293T (*see* **Note 18**), and place the cells at 37 °C in a 5% humidified CO_2 incubator.
2. About 1 h, prior to viral transduction, thaw 1.5 mL of the LV on ice if previously stored at −80 °C. Briefly spin down the vial before opening it, under the hood.
3. At 5 min to infection, sanitize the vial containing the virus with 70% ethanol and place it in the hood. Pick up 0.5 mL of the LV stock into a sterile 1.5 mL tube and top it up to 1 mL with OptiMEM. Add to both LV stocks 8 μg/mL of polybrene (Hexadimethrine bromide) (*see* **Note 19**), swirl and rock the vial to mix, without vortexing.
4. Take the incubated cells, discard the old medium, and then add the virus-polybrene mix to the cells. Incubate the plate at 37 °C, 5% CO_2.
5. Rock and tilt the plate back and forth, every 20 min for at least the first 80 min of incubation (*see* **Note 20**).
6. Incubate the cell-viral transduction culture for 6 h and then adjust the medium to 2 mL with a fresh complete DMEM for Lenti-X 293T. Incubate overnight, at 37 °C, 5% CO_2.
7. 16–20 h post-infection, change the medium to a fresh one without polybrene addition and continue the incubation, at 37 °C, 5% CO_2, for the next 24 h.
8. Select cells with puromycin, 48 h post-transduction. Replace the medium on both the positive and control cultures with complete DMEM containing puromycin, 1 μg/mL (*see* **Note 21**). Incubate the plate at 37 °C, 5% CO_2. Observe cell death daily. The control plate should have complete cell death by the expected days, postpuromycin treatment. By eye observation, compare the difference in cell death between the 0.5 and 1 mL

virus supernatant relative to the control, to have a fair idea of the virulence of the produced LV stock. To estimate the transduction efficiency, continue with the steps below.

9. Carefully wash cells with 1× PBS to remove completely all dead cells postpuromycin treatment. Add fresh complete DMEM and continue to grow cells at 37 °C, 5% CO_2.
10. Between 16 and 24 h post-transduction, estimate the transduction efficiency by a reporter gene such as GCaMP6s expression analysis from the transfer vector, if the GCaMP6s is constitutively expressed. If inducible under the tight TRE-promoter (Fig. 1), add 2 μg/mL dox to the cell culture medium, rock and swell the plate to evenly mix and distribute the dox, to activate the GCaMP6s expression (*see* **Note 22**).
11. Determine GCaMP6s expression with a fluorescent microscope or other techniques, 24 h after dox addition to the cell culture. The level of GCaMP6s expression can confirm the strength of the produced virus titer.

3.4 Establishment of Stable HeLa DH Cx32 Transgenic Cell Lines (Fig. 3)

1. One day before viral infection, seed two 1.0×10^5 HeLa DH cells in a 35 mm plate in a complete DMEM (high glucose 4.5 g/L, pyruvate) with 10% TSA-FBS (*see* **Note 2**), 2 mM glutamine, and 1% penicillin/streptomycin, for HeLa cells. Place the cells at 37 °C in a 5% humidified CO_2 incubator for 16–20 h.
2. 1 h, prior to viral transduction, thaw 1 mL of the LV on ice or use a freshly harvested virus. The transducing titer is established from the virus viability studies.
3. About 5 min to infection, briefly spin down the vial containing the virus, sanitize it with 70% ethanol, and place it in the hood. Add 8 μg/mL of polybrene (Hexadimethrine bromide) (*see* **Note 19**), and rock the vial to mix, without vortexing.
4. Pick the overnight incubated cells, discard the old medium, and then add the virus-polybrene mix to the cells. Return the plate to the incubator, at 37 °C, 5% CO_2.
5. Gently swirl and tilt the plate back and forth, every 20 min for at least the first 80 min of incubation, to help spread the virus uniformly to the cell monolayer. This enhances higher transduction efficiency. Incubate the cell-viral culture for 5 h.
6. Adjust the medium to 2 mL with a fresh complete DMEM for HeLa DH cells. Incubate the plate overnight, at 37 °C, 5% CO_2.
7. 16 h post-infection, replace the medium with a fresh one without polybrene. Continue the incubation, at 37 °C, 5% CO_2, for the next 24 h.

8. Select positive HeLa DH cells with 1 μg/mL of puromycin, 48 h post-transduction, in complete DMEM for HeLA DH cells. Incubate the plate at 37 °C, 5% CO_2. Observe cell death the next day (*see* **Note 23**).
9. Carefully wash the transduced HeLA DH cells with 1× PBS to remove dead cells postpuromycin treatment. Add fresh complete DMEM for HeLa DH cells, and continue to grow cells at 37 °C, 5% CO_2. Incubate for a further 16–24 h and then proceed to either step 10 (induction of GOI for downstream applications; expressed genes detection and functional assay analysis) or 11 (expand and freeze stable HeLa DH cells for future use).
10. Add 4 μg/mL dox to the cell culture medium, rock, and swell the plate to evenly mix and distribute the dox, to activate the TRE-promoter for the Cx32-linked GCaMP6s- reporter expression (*see* **Note 24**).
11. For stable HeLa DH cell maintenance, passage the transduced cells in a sterile 100 mm plate in a fresh complete DMEM for HeLa DH cells. At about 80% confluence, split and expand the cells using a 100 mm plate. Passage the cells two more times and proceed to freezing (*see* **Note 25**; *see* Fig. 3).

3.5 Freezing and Storing HeLA DH Cx32 Transgenic Cell Lines

1. Freeze the HeLa DH cells, at 80% confluence.
2. Prewarm 1× PBS, trypsin-EDTA, and complete DMEM for HeLa DH at 37 °C. Under the hood, wash cells with 5 mL 1× PBS, and detach the cells with 1 mL trypsin-EDTA treatment.
3. Incubate the plate for 3 min, collect the cells with complete DMEM for HeLa DH, and pellet the cells at 250 × *g* for 3 min.
4. Resuspend the cells in 900 μL of the complete medium by up-and-down pipetting ten times.
5. Add 100 μL of sterile DMSO, and do up-and-down pipetting twice to mix.
6. Aliquot 500 μL of the cell suspension to a detailed label cryovial containing necessary information: name of cell line, gene, plasmid name, date, and any other useful information.
7. Immediately, transfer the vial to ice and then transfer it into a freezing box for storage at −80 °C, for a few days. For long-term storage, transfer the vials to liquid nitrogen tanks.

3.6 Retrieving Stable HeLA DH Cx32 Transgenic Cell Lines from Frozen Stocks

1. Warm complete HeLa DH medium.
2. Pick the frozen cell vial from storage and thaw it rapidly at 37 °C in a water bath. Wipe the vial with 70% ethanol to sanitize it and place it in the hood. In a sterile 15 mL conical centrifuge tube containing 4.5 mL of the prewarmed complete medium, gently transfer the thawed cell content.

3. Pellet cells for 3 min at 250 × *g*. Gently aspirate the supernatant leaving the cell pellet untouched. Add 1 mL of medium and gently resuspend the cells by up-and-down pipetting just <5 times. This step allows the complete removal of the DMSO used in cryopreserving the cells during the freezing.
4. Top up to 10 mL of the cell suspension with fresh complete medium and transfer it into a sterile 100 mm plate. Gently swirl the plate to evenly distribute the cells over the growth area. Gently place the plates in a humidified incubator at 37 °C, 5% CO_2.
5. The next day, examine the characteristics of the plated cells under the microscope. Change the medium to a prewarm fresh complete growth medium. Continue the incubation and split the cells the next day (*see* **Note 26**).
6. Prepare extra plates for freezing, if necessary.

3.7 Immunolabeling of HeLa DH Cx32 Transgenic Cells In Vivo

1. Plate 1 × 10^5cells in two Petri dishes (35 mm), each containing five coverslips (12 mm diameter) in complete HeLa DH growth medium and let cell grow for 24 h at 37 °C, 5% CO_2 humidified incubator (*see* **Note 27**).
2. Prepare doxycycline complete HeLa DH growth medium.
3. Aspirate the growing medium from the dishes and replace with complete medium alone or with doxycycline and incubate for 72 h.
4. Aspirate the growth medium from the dishes.
5. Transfer each coverslip to a single appropriately labeled well of a 24-well plate, containing 200 μL of abEC1.1hIgG solution or only 200 μL of diluting solution (*see* **Note 28**) and incubate for 1 h.
6. Aspirate the solutions, transfer the 24-well plate on ice and wash each coverslip once with ice cold 1× PBS (*see* **Note 29**).
7. Incubate in fixing solution for 10 min at room temperature.
8. Prepare secondary antibody dilution using Eppendorf tubes, as follows: Alexa Fluor 594-conjugated Goat anti-Human IgG (1:800, Jackson Immuno Research, Cat. # 109-585-170). Protect diluted secondary antibodies from light and store dilutions at 4 °C until ready to use (*see* **Note 6**).
9. Aspirate fixing solution and wash three times with 1× PBS.
10. Add the secondary antibody solution to each coverslip, insert the well in the humidity chamber and incubate 1 h, at room temperature.
11. Aspirate the secondary antibody solution and wash coverslips three times with 1× PBS.

12. Permeabilize fixed cells with permeabilization solution for 5 min, at room temperature (*see* **Note 30**).
13. Aspirate the permeabilization solution and wash coverslips once with 1× PBS.
14. Add blocking solution to each coverslip and incubate for 1 h at room temperature.
15. While blocking, prepare the primary antibody dilution: in this case Rabbit polyclonal anti-green fluorescent protein-GFP (1: 200, Invitrogen, Cat. # A-11122).
16. Aspirate blocking solution.
17. Add the primary antibody solution to each coverslip (*see* **Note 31**), insert the well in the humidity chamber and incubate over-night, at 4 °C.
18. Repeat **steps 9–12** for secondary antibody staining, using Alexa Fluor 488-conjugated Donkey anti-Rabbit IgG (1:800, Invitrogen, Cat. # A-21206).
19. To stain all cell nuclei, add DAPI solution to each coverslip and incubate for 10 min at room temperature.
20. Aspirate DAPI solution and wash three times with 1× PBS and once with ddH_2O.
21. Label glass microscope slides with the date, cell line, transfected plasmid (if applicable), immunolabeled proteins and secondary antibodies.
22. Add 1 drop of ProLong™ Gold mounting medium on the glass microscope slide for each coverslip (*see* **Note 32**). Ensure the drops of mounting media are spaced out sufficiently to fit the coverslips (*see* **Note 33**).
23. Using forceps, carefully lift each coverslip from the well and gently dab the side of the coverslip onto a piece of filter paper to remove excess ddH_2O.
24. Place the coverslip cell-side-down onto the drop of mounting medium. Invert slide with coverslip onto paper towel and gently press down to remove excess fluid.
25. Place glass microscope slides coverslip-side-up on a flat dry surface in the dark at room temperature for at least 2 h (*see* **Note 34**) and then transfer to a slide folder and store at 4 °C. Prepared slides maintained in the dark can be imaged many days after preparation with minimal photobleaching (*see* Fig. 4).

3.8 Immunolabeling of Fixed and Permeabilization and HeLa DH Cx32 Transgenic Cells

1. Plate 1×10^5cells in two Petri dishes (35 mm), each containing five coverslips (12 mm diameter) in growth medium and let cell grow for 24 h at 37 °C, 5% CO_2 humidified incubator (*see* **Note 27**).
2. Prepare doxycycline growth medium.

3. Aspirate the growth medium from the dishes and replace with growth medium alone or doxycycline growth medium and incubate for 72 h.
4. Aspirate the growth medium from the dish.
5. Gently rinse the cell monolayer once with 2 mL of sterile 1× PBS.
6. Prepare a 24-well plate containing 200 μL of 1× PBS in a number of wells corresponding to the number of coverslips (*see* **Note 28**).
7. Transfer each coverslip to a single appropriately labeled well of a 24-well plate and then aspirate 1× PBS.
8. Incubate in fixing solution for 10 min at room temperature, on a rocking platform.
9. Aspirate fixing solution and wash once with 1× PBS (*see* **Note 35**).
10. Permeabilize fixed cells with permeabilization buffer for 5 min, at room temperature.
11. Add blocking solution (*see* **Note 36**) to each coverslip and incubate for 1 h at room temperature.
12. While blocking, prepare the primary antibody dilution in blocking solution as per the antibody information sheet or from experiential learning as to the most appropriate concentration or dilution to use. Here we use mouse monoclonal anti-Cx32 (1:100, Invitrogen, Cat. # 35–8900, clone 5F9A9).
13. Prepare a lightproof humidity chamber using a dark plastic box with lid. Cut a piece of Whatman filter paper to fit inside the box and soak with ddH_2O. Cover with the lid.
14. Aspirate blocking solution.
15. Add the first primary antibody solution to each coverslip, insert the well in the humidity chamber and incubate over-night, at 4 °C (*see* **Note 31**).
16. After first primary antibody incubation, prepare corresponding secondary antibody dilution using 1.5 mL Eppendorf tube, as follows: Alexa Fluor 555-conjugated Donkey anti-Mouse IgG (1:800, Invitrogen, Cat. # A-31570). Protect diluted secondary antibodies from light and store dilutions at 4 °C until ready to use (*see* **Note 6**).
17. Aspirate the primary antibody solution and wash coverslips three times with 1× PBS.
18. Add the first secondary antibody solution to each coverslip, insert the well in the humidity chamber and incubate 1 h, at room temperature.

19. Aspirate the secondary antibody solution and wash coverslips three times with 1× PBS.
20. To double label, repeat **steps 15–19** with the second set of primary and secondary antibody solutions (i.e., and Rabbit polyclonal anti-green fluorescent protein-GFP, 1:200, Invitrogen, Cat. # A-11122, and Alexa Fluor 488-conjugated Donkey anti-Rabbit IgG, 1:800, Invitrogen, Cat. # A-21206).
21. Add DAPI solution to each coverslip and incubate for 10 min at room temperature to stain the nuclei.
22. Aspirate DAPI solution and wash three times with 1× PBS and once with ddH_2O.
23. Label glass microscope slides with the date, cell line, transfected plasmid (if applicable), immunolabeled proteins and secondary antibodies.
24. Add 1 drop of ProLong™ Gold mounting medium on the glass microscope slide for each coverslip (*see* **Note 32**). Ensure the drops of mounting media are spaced out sufficiently to fit the coverslips (*see* **Note 33**).
25. Using forceps, carefully lift the coverslip from the well and gently dab the side of the coverslip onto a piece of filter paper to remove excess ddH_2O.
26. Place the coverslip cell-side-down onto the drop of mounting medium. Invert slide with coverslip onto paper towel and gently press down to remove excess fluid.
27. Place glass microscope slides coverslip-side-up on a flat dry surface in the dark at room temperature for at least 2 h (*see* **Note 34**) and then transfer to a slide folder and store at 4°C. Prepared slides maintained in the dark can be imaged many days after preparation with minimal photobleaching (*see* Fig. 5).

3.9 Imaging

Here, we describe how immunofluorescence images can be obtained on a Leica TCS SP5 confocal microscope.

1. Image acquisition is achieved using the Leica Microsystems Leica Application Suite Advanced Fluorescence 2.7.3.9723 (LAS AF) software platform installed on a Leica TCS SP5 imaging system.
2. Using the LAS AF software, open Configuration tab and select Laser configuration to switch on 405-Diode, Argon and helium–neon (HeNe) 543. Set Argon power around 15% of the maximum laser power.
3. Open Acquire tab and set the acquisition parameters in the Acquisition tab. Set the Scan sequence as follows (*see* **Note 37**):

(a) Scan 1: DAPI, excited by 405-nm Diode laser (intensity 8%) and collected between 441 and 487 nm.

(b) Scan 2: Alexa Fluor 488, excited by 488-nm Argon laser (intensity 15%) and collected between 507 and 556 nm.

(c) Scan 3: Alexa Fluor 555, excited by 543-nm HeNe laser (intensity 30%) and collected between 561 and 623 nm or Alexa Fluor 594, excited by 543-nm HeNe laser (intensity 46%) and collected between 585 and 706 nm.

Set "between stacks" acquisition mode, that allows to acquire each channel sequentially (*see* **Note 38**).

4. On the microscope, select the proper oil immersion objective lens (*see* **Note 39**).
5. Tilt the microscope head back to access the objective lenses and apply one drop of immersion oil onto the objective.
6. Place the glass microscope slide gently onto the mechanical stage with the coverslip directed above the objective lens. Use the coarse and fine adjustment knob to raise the lens objective into contact with the immersion oil. Return the microscope head to the upright position.
7. Turn on the fluorescence lamp to focus the sample and check the fluorescence of the different fluorochromes.
8. While looking through the ocular lenses, focus on the specimen and use the joystick controller to orient the field of view and locate a region of interest (ROI).
9. Using the LAS AF software, under the acquisition tab, select the live button (*see* **Note 40**). Adjust the fine focus to emphasize connexins found in subcellular organelles, vesicles, cell surface distributions, and gap junction plaques.
10. Return to the channel window and adjust the laser strength to minimize pixel saturation (*see* **Note 41**) and select optimal frame size and master gain for image acquisition.
11. To acquire an image, select the Start button under the acquisition tab (*see* **Note 42**).
12. For the acquisition, we typically select an image dimension of either 1024 × 1024 pixels or 2048 × 2048 pixels to have higher resolution.
13. Export single or multiple images as a BMP, TIFF, PNG, WDP, or SUR file (*see* **Note 43**). Export all channel collected images to the desired destination folder.
14. To acquire a z-stack, under the Acquisition tab, check the z-stacking box, advance to the z-stack window and select the live button. Then use the fine focus to capture the upper boundary of the cell as defined by evidence of diminishing fluorescence. Select the set Begin button.

15. Adjust the focus to identify the opposite boundary (base) of the cell where minimal fluorescence will be detected. Select the End button.
16. Adjust the number of optical slices to be captured (*see* **Note 44**).
17. Under the acquisition tab, unselect the Live button then select Start experiment.
18. Export single or multiple images, as described at step 11.
19. Z-section acquired images are stacked for visualization purposes. Click on Process and select 3D Projection in Tools tab. Using Maximum Method option from the dropdown list.
20. Using the extended depth of focus or orthogonal projection processing features, both found within the LAS AF software, using Maximum Projection option from the method dropdown list and export as TIFF in the destination folder.
21. Save the project (*see* **Note 45**).

4 Notes

1. Strict safety and precautions should be followed, to prevent contamination of cell culture and protection from the recombinant LV production. For LV production and infection, Biosafety level 2 rules should be applied.
2. Use of prewarmed solutions in preparing cell cultures reduces osmotic shock. It is recommended to use prewarmed cell culture medium when thawing cells from frozen stock to enhance cell survival.
3. Increase LV titers using heat-inactivated tetracycline-free (Tet-System Approved) FBS. The normal FBS are known to be contaminated with traces of tetracycline or its derivatives. Due to the activation of GOI expression under the tetracycline-inducible promoter (tight TRE promoter, Fig. 1), traces of tetracycline can interfere with the virus protein expression, leading to low viral titer. Heat-inactivation of FBS provides a means to inactivate complement proteins present in the FBS, reducing any unwanted interactions with the packaging virus or host cell proteome, and during immunological assays [33]. Heat-inactivated FBS is prepared by incubating completely thawed FBS for 30 min at 56 °C submerged-water bath, with gentle shaking and then cooling on ice. Aliquots of 50 mL can be prepared and frozen for future use.
4. Doxycycline is light sensitive, therefore care should be taken when handling it, for example by covering the containers with tin foil. Once thawed, this solution can be stored at +4 °C for a few months.

5. May need to vortex for several minutes to completely dissolve the BSA powder. It is recommended to prepare it fresh for every experiment.
6. Immediately before use, carefully centrifuge the secondary antibody dilutions using a bench Eppendorf centrifuge at 14,000×g for 1–2 min at room temperature to remove any possible impurity from the solution that may affect confocal microscopy imaging. Do not perturb the pellet while collecting the necessary antibody amount. From this step on it is recommended to protect the wells from light.
7. 1xDAPI solution in 1× PBS should be freshly prepared immediately before use and kept in the dark at room temperature.
8. The methods described here to image and localize connexins in fixed cells were originally developed and refined from our experience with our current instrument. It should be possible to adapt the methods described here to all currently used confocal microscope systems.
9. It is recommended to use a cell line that stably expresses the SV40 large T antigen such as HEK 293T to facilitate high viral titer production. Using a cell line without the SV40 large T antigen expression can result in lower lentiviral titers.
10. Lenti-X 293T like other HEK293T cells detach easily from the culture plate, therefore the poly-L-lysine coats help immobilize the cells to the plate and prevent easy cell detachment when transfecting or replacing the culture medium.
11. After up-and-down pipetting, it is recommended not to further rock the plate and handle it gently while moving it into the incubator, to maintain uniform cell distribution in the plate. A nonuniform cell distribution affects the transfection efficiency, leading to low lentiviral titer.
12. To reduce up and down pipetting of the Opti-MEM/Lipo3000 mix, which could reduce transfection efficiency, always add the DNA mix in tube 1 to the lipofectamine 3000 mix in tube 2. For accurate pipetting, adjust the tube 1 mix to 650 μL and pick the exact volume of 600 μL. Discard the remaining volume. Longer incubation of the transfection mix to up to 35 min, beyond the 15 min suggested in the reagent datasheet enhances the transfection efficiency for high titer virus production.
13. OptiMEM I with Reduce Serum is optimal for transient cell transfection. However, failure to replace the transfection medium with complete DMEM after 5 h post-transfection could weaken cell detachment, due to the reduced serum, leading to cell death and low transfection efficiency. Also, the complete DMEM for Lenti-X 293T has high supplements and

the right pH to support lentiviral particles production, which starts >13 h post-transfection.

14. Henceforth, the culture could contain a trace amount of the LV proteins and particles, so extreme care must be taken to avoid direct personal and environmental contact with the culture, without prior treatment of the used tips, serological pipettes, and the discarded solutions in 20% bleach.
15. For a high titer LV production, high transfection efficiency is required. Observed green, fluorescent positive cells should be >75% to 90%. A percentage of fluorescent cells <60% indicates an inadequate transfection for LV production.
16. It is recommended to use a PES filter instead of a nitrocellulose filter, which can potentially bind the virus surface proteins, reducing virus titers. Although a 0.20 μm PES filter can be used, 0.45 μm is preferred since the pore size is bigger enough to reduce the clogging of the filter pores by cell debris remaining in the collected supernatant after the centrifugation. Failure to filter the supernatant can lead to virus producing 293T cell contaminations of the lentiviral preps.
17. The aliquot can be stored temporarily on ice or at 4 °C. Avoid repeated cycles of LV freezing and thawing to maintain the virus titer as high as possible. Depending on the transfer gene size, if >80% transfection efficiency is observed, the first LV collection should give a strong viral titer, of $>10^6$ to 10^8 transducing units per mL (TU/mL) when measured. As such, 1 mL aliquots of the first harvested LV stock are enough to transduce naïve targeted cells. Even if the targeted cells require a separate growth medium, transduce the cells with the LV stock for 2–5 h, then replace the stock medium with the right growth medium. For naïve cells that are difficult to infect, a higher viral titer is required; it is recommended to concentrate the virus by ultracentrifugation techniques, or by using an appropriate kit to concentrate the virus in 1× PBS or OptiMEM (not detailed in this protocol).
18. Adjust the seeded cell density at about 50–60% confluency at the time of viral transduction based on predetermined targeted cell viability and doubling capacity, as well as the number of days required to complete downstream analysis.
19. For most mammalian cells, 5–10 μg/mL polybrene can be used to neutralize the charges on the virus and cell surfaces, reducing charge repulsion and thus enhancing transduction. It is recommended to first determine the cell polybrene sensitivity limit by adding varying concentrations (2–10 μg/mL) to the target cells. Secondly, perform viral transduction efficiency with varying polybrene concentrations to determine the best titer concentration. For cells sensitive to polybrene, use lower

concentrations, limit the transduction time to 5–8 h, and replace the transduction mix with a fresh complete medium.

20. This procedure helps spreading the virus uniformly to the cell monolayer, leading to higher transduction efficiency.
21. It is recommended to obtain a puromycin-killing curve to determine the right dosage needed to kill the nonstable cells, not expressing the transfer gene resistant to puromycin. To this end, plate the same number of cells to be used for stable line generation. On day 2, change the cell medium and add varying amounts of puromycin, 1–10 μg/mL to the plated cells. Observe cell viability daily, for at least 2 days. Choose the minimum amount of puromycin that can result in cell death by day 2. Use this puromycin concentration to select a stable cell line expressing the transgene of choice.
22. This step is only required when the transfer gene reporter such as GCaMP6s needs to be expressed under the doxycycline inducible promoter. This is used to estimate the transduction efficiency by eye, FACs, fluorescent gene quantification by qRT-PCR or WB, and an estimate of the multiplicity of infection is then determined.
23. It is expected that the nonstable HeLa DH cells in the pool will all be dead by days 2–3 of the puromycin treatment. Further propagation of previously selected cells in a medium containing puromycin can lead to monoclonal stable cell selection.
24. Doxycycline can be toxic to cells at a concentration >1 μg/mL. It is recommended to perform a dosage curve (1–10 μg/mL) to determine toleration limit of your target cells. For high dosage, a 1–2 μg/mL can be added gradually, every 24 h to enhance the TRE-promoter activation, increasing the targeted gene expression. For toxic transgene expression, add the minimum dosage and for a short predetermined incubation period.
25. For toxic gene expression under the inducible TRE-promoter, expand the HeLa DH cells and only activate the TRE-promoter when enough stable cell lines have been expanded and stored frozen, at −80 °C.
26. Passage the cells at least once more. If the cells appear to grow normally, then they are ready further use, such as gene expression activation for molecular and functional analysis.
27. It is good practice to include extra coverslips in each dish in the event a coverslip becomes damaged, or a late decision is made to expand the number of antibodies used for immunolabeling.
28. To ensure standardization, all the incubations from now on will be performed using 200 μL of any type of solutions in a 24-well plate always covered by the lid, and on a rocking platform.

29. This step is performed on ice to block binding of the antibody to the antigens and to stop cell growth.
30. As abEC1.1hIgG recognizes antigens located on the extracellular domain of Cx32 protein, permeabilization is performed only after abEC1.1hIgG incubation.
31. It is good practice to always include control samples, including coverslips incubated only with the dilution solution of the primary antibody (without the latter), but incubated with the appropriate secondary antibody.
32. We normally use ProLong™ Gold as mounting medium to increase fluorophore photostability.
33. Up to four 12 mm glass coverslips can easily fit on a standard glass microscope slide.
34. Curing time for Prolong Gold may vary from a couple of hours to overnight, depending on the thickness of the sample and the relative humidity of the surrounding air.
35. Fixed cells can be stored at 4 °C submerged in 1× PBS for up to 2 weeks, although optimal results are achieved if cells are immunolabeled shortly after fixation.
36. Use 2% BSA instead of 0.5% to reduce anti-Cx32 antibody nonspecific binding.
37. Laser intensity should be kept to a minimum to avoid fluorochrome photobleaching. In cases where images are to be quantified and compared within an experimental set, identical settings need to be used to acquire all images.
38. Most imaging systems can be programmed to have several preset imaging protocols that activate the desired lasers and set the image acquisition parameters for each detector. If a protocol is already present, upload the protocol and skip point 3.
39. In the case of Figs. 4 and 5, we used either 40×/1.25 NA or 63×/1.4 NA oil immersion objective. In general, the use of a high magnification objective, such as the HC PL Apo 63×/1.4 NA oil immersion objective, enables the capture of high-resolution images that clearly resolve connexins in multiple subcellular locations.
40. We typically set the acquisition speed at 400 Hz and the image dimension at 512 × 512 pixels for the live imaging option that allows faster visualization and less photodamage while selecting the field of view to be acquired.
41. It is important to ensure images are not saturated. The laser strength and master gain can be manually adjusted to minimize pixel saturation. To visualize pixel saturation, select the QLUT button, to change the colors to intensity values. Saturated

pixels will appear blue. Aim for minimal pixel saturation by ensuring only the regions with the highest signal appear blue/saturated.

42. As an example of parameters commonly used, image dimension of 1024 × 1024 pixels or 2048 × 2048 pixels, a scan speed of 400 Hz, line averaging between 3 and 6 in each and a zoom between 3 and 5. Conditions for image capture are customized depending on the strength of the fluorescent signals and the desired signal-to-noise ratio. For example, increasing averaging greatly improves the signal-to-noise ratio.
43. We typically export images as TIFF files.
44. For monolayer cultures of HeLa cells, we typically collect three to five images, where each optical section captures ~0.5 μm of optical fluorescence.
45. The project containing all the images acquired and the corresponding metadata are saved in a file in Leica Image Format (LIF).

Acknowledgments

The authors thank Chiara Peres for expert assistance with confocal microscopy. pMD2.G (Addgene plasmid #12259; http://n2t.net/addgene:12259) and psPAX2 (Addgene plasmid #12260 SVC; http://n2t.net/addgene:12260; RRID: Addgene_12260), were a kind gift from Didier Trono. This work was funded by grants awarded to Fabio Mammano by Fondazione Telethon (GGP19148) and the Italian Ministry of Education, Universities and Research (PRIN 20175C22WM).

References

1. Elfgang C, Eckert R, Lichtenberg-Frate H et al (1995) Specific permeability and selective formation of gap junction channels in connexin-transfected HeLa cells. J Cell Biol 129(3): 805–817. https://doi.org/10.1083/jcb.129.3.805
2. Beltramello M, Bicego M, Piazza V et al (2003) Permeability and gating properties of human connexins 26 and 30 expressed in HeLa cells. Biochem Biophys Res Commun 305(4): 1024–1033. https://doi.org/10.1016/s0006-291x(03)00868-4
3. Beltramello M, Piazza V, Bukauskas FF et al (2005) Impaired permeability to Ins(1,4,5)P3 in a mutant connexin underlies recessive hereditary deafness. Nat Cell Biol 7(1):63–69. https://doi.org/10.1038/ncb1205
4. Vargas AA, Cisterna BA, Saavedra-Leiva F et al (2017) On biophysical properties and sensitivity to gap junction blockers of connexin 39 hemichannels expressed in HeLa cells. Front Physiol 8:38. https://doi.org/10.3389/fphys.2017.00038
5. Choi EJ, Palacios-Prado N, Saez JC et al (2020) Identification of Cx45 as a major component of GJs in HeLa cells. Biomol Ther 10(10). https://doi.org/10.3390/biom10101389
6. Panosyan FB, Laura M, Rossor AM et al (2017) Cross-sectional analysis of a large cohort with X-linked Charcot-Marie-Tooth disease (CMTX1). Neurology 89(9): 927–935. https://doi.org/10.1212/WNL.0000000000004296

7. Nardin C, Tettey-Matey A, Donati V et al (2022) A quantitative assay for Ca2+ uptake through normal and pathological hemichannels. Int J Mol Sci 23(13):7337. https://doi.org/10.3390/ijms23137337
8. Malim MH, Hauber J, Le SY et al (1989) The HIV-1 rev trans-activator acts through a structured target sequence to activate nuclear export of unspliced viral mRNA. Nature 338(6212):254–257. https://doi.org/10.1038/338254a0
9. Naldini L, Blomer U, Gallay P et al (1996) In vivo gene delivery and stable transduction of nondividing cells by a lentiviral vector. Science 272(5259):263–267. https://doi.org/10.1126/science.272.5259.263
10. Zufferey R, Nagy D, Mandel RJ et al (1997) Multiply attenuated lentiviral vector achieves efficient gene delivery in vivo. Nat Biotechnol 15(9):871–875. https://doi.org/10.1038/nbt0997-871
11. Dull T, Zufferey R, Kelly M et al (1998) A third-generation lentivirus vector with a conditional packaging system. J Virol 72(11): 8463–8471. https://doi.org/10.1128/JVI.72.11.8463-8471.1998
12. Zennou V, Petit C, Guetard D et al (2000) HIV-1 genome nuclear import is mediated by a central DNA flap. Cell 101(2):173–185. https://doi.org/10.1016/S0092-8674(00)80828-4
13. Chen TW, Wardill TJ, Sun Y et al (2013) Ultrasensitive fluorescent proteins for imaging neuronal activity. Nature 499(7458):295–300. https://doi.org/10.1038/nature12354
14. Gossen M, Bujard H (1992) Tight control of gene expression in mammalian cells by tetracycline-responsive promoters. Proc Natl Acad Sci USA 89(12):5547–5551. https://doi.org/10.1073/pnas.89.12.5547
15. Fallon AM, Stollar V (1982) Isolation and characterization of puromycin-resistant clones from cultured mosquito cells. Somatic Cell Genet 8(4):521–532. https://doi.org/10.1007/BF01538712
16. Essenfelder GM, Bruzzone R, Lamartine J et al (2004) Connexin30 mutations responsible for hidrotic ectodermal dysplasia cause abnormal hemichannel activity. Hum Mol Genet 13(16):1703–1714. https://doi.org/10.1093/hmg/ddh191
17. Stong BC, Chang Q, Ahmad S et al (2006) A novel mechanism for connexin 26 mutation linked deafness: cell death caused by leaky gap junction hemichannels. Laryngoscope 116(12):2205–2210. https://doi.org/10.1097/01.mlg.0000241944.77192.d2
18. Lee JR, Derosa AM, White TW (2009) Connexin mutations causing skin disease and deafness increase hemichannel activity and cell death when expressed in Xenopus oocytes. J Invest Dermatol 129(4):870–878. https://doi.org/10.1038/jid.2008.335
19. Sanchez HA, Mese G, Srinivas M et al (2010) Differentially altered Ca2+ regulation and Ca2+ permeability in Cx26 hemichannels formed by the A40V and G45E mutations that cause keratitis ichthyosis deafness syndrome. J Gen Physiol 136(1):47–62. https://doi.org/10.1085/jgp.201010433
20. Terrinoni A, Codispoti A, Serra V et al (2010) Connexin 26 (GJB2) mutations, causing KID syndrome, are associated with cell death due to calcium gating deregulation. Biochem Biophys Res Commun 394(4):909–914. https://doi.org/10.1016/j.bbrc.2010.03.073
21. Sanchez HA, Villone K, Srinivas M et al (2013) The D50N mutation and syndromic deafness: altered Cx26 hemichannel properties caused by effects on the pore and intersubunit interactions. J Gen Physiol 142(1):3–22. https://doi.org/10.1085/jgp.201310962
22. Berger AC, Kelly JJ, Lajoie P et al (2014) Mutations in Cx30 that are linked to skin disease and non-syndromic hearing loss exhibit several distinct cellular pathologies. J Cell Sci 127(Pt 8):1751–1764. https://doi.org/10.1242/jcs.138230
23. Lu Y, Zhang R, Wang Z et al (2018) Mechanistic effect of the human GJB6 gene and its mutations in HaCaT cell proliferation and apoptosis. Braz J Med Biol Res 51(9):e7560. https://doi.org/10.1590/1414-431X20187560
24. Bolivar F, Rodriguez RL, Greene PJ et al (1977) Construction and characterization of new cloning vehicles. II. A multipurpose cloning system. Gene 2(2):95–113
25. Castagnoli L, Scarpa M, Kokkinidis M et al (1989) Genetic and structural analysis of the ColE1 Rop (Rom) protein. EMBO J 8(2): 621–629. https://doi.org/10.1002/j.1460-2075.1989.tb03417.x
26. Hurwitz J, Dean FB, Kwong AD et al (1990) The in vitro replication of DNA containing the SV40 origin. J Biol Chem 265(30): 18043–18046
27. Nardin C, Peres C, Putti S et al (2021) Connexin hemichannel activation by S-nitrosoglutathione synergizes strongly with photodynamic therapy potentiating antitumor bystander killing. Cancers 13(20). https://doi.org/10.3390/cancers13205062

28. Xu L, Carrer A, Zonta F et al (2017) Design and characterization of a human monoclonal antibody that modulates mutant connexin 26 hemichannels implicated in deafness and skin disorders. Front Mol Neurosci 10:298. https://doi.org/10.3389/fnmol.2017.00298
29. Ziraldo G, Buratto D, Kuang Y et al (2019) A human-derived monoclonal antibody targeting extracellular connexin domain selectively modulates hemichannel function. Front Physiol 10: 392. https://doi.org/10.3389/fphys.2019.00392
30. Kuang Y, Zorzi V, Buratto D et al (2020) A potent antagonist antibody targeting connexin hemichannels alleviates Clouston syndrome symptoms in mutant mice. EBioMedicine 57: 102825. https://doi.org/10.1016/j.ebiom.2020.102825
31. Peres C, Sellitto C, Nardin C et al (2023) Antibody gene transfer treatment drastically improves epidermal pathology in a keratitis ichthyosis deafness syndrome model using male mice. EBioMedicine 89:104453. https://doi.org/10.1016/j.ebiom.2023.104453
32. Buratto D, Donati V, Zonta F et al (2021) Harnessing the therapeutic potential of antibodies targeting connexin hemichannels. Biochim Biophys Acta Mol Basis Dis 1867(4): 166047. https://doi.org/10.1016/j.bbadis.2020.166047
33. Tate J, Ward G (2004) Interferences in immunoassay. Clin Biochem Rev 25(2):105–120

Chapter 13

A Protocol for the Automated Assessment of Cutaneous Pathology in a Mouse Model of Hemichannel Dysfunction

Chiara Peres and Fabio Mammano

Abstract

In this chapter, we provide detailed instructions to perform quantitative reflectance imaging in a mouse model of a rare epidermal disorder caused by hyperactive connexin 26 hemichannels. Reflectance imaging is a versatile and powerful tool in dermatology, offering noninvasive, high-resolution insights into skin pathology, which is essential for both clinical practice and research. This approach offers several advantages and applications. Unlike traditional biopsy, reflectance imaging is noninvasive, allowing for real-time, in vivo examination of the skin. This is particularly valuable for monitoring chronic conditions or assessing the efficacy of treatments over time, enabling the detailed examination of skin morphology. This is crucial for identifying features of skin diseases such as cancers, inflammatory conditions, and infections. In therapeutic applications, reflectance imaging can be used to monitor the response of skin lesions to treatments. It can help in identifying the most representative area of a lesion for biopsy, thereby increasing the diagnostic accuracy. Reflectance imaging can also be used to diagnose and monitor inflammatory skin diseases, like psoriasis and eczema, by visualizing changes in skin structure and cellular infiltration. As the technology becomes more accessible, it has potential in telemedicine, allowing for remote diagnosis and monitoring of skin conditions. In academic settings, reflectance imaging can be a powerful research tool, enabling the study of skin pathology and the effects of novel treatments, including the development of monoclonal antibodies for therapeutic applications.

Key words Keratitis-ichthyosis-deafness syndrome, Connexin 26, Hemichannels, Optics, System automation, Reflectance imaging, Feature detection

1 Introduction

Genodermatoses are inherited skin diseases that often present with multisystem involvement leading to increased morbidity and mortality [1]. A group of rare genodermatoses has been causally linked to connexin (Cx) mutations [2]. Increasing evidence points to leaky or hyperactive HCs as a common gain-of-function etiopathogenetic mechanism for a subgroup of genodermatoses with autosomal dominant inheritance and overlapping spectrum of cutaneous clinical features, linked to a subset of Cx26, Cx30, and Cx43

Fabio Mammano and Mauricio Retamal (eds.), *Connexin Hemichannels: Methods and Protocols*, Methods in Molecular Biology, vol. 2801, https://doi.org/10.1007/978-1-0716-3842-2_13,

mutations, which may drastically compromise quality of life and life-long health [3, 4]. Among these, keratitis-ichthyosis-deafness (KID) syndrome is a dysregulated hemichannel disorder that causes skin abnormalities, eye problems, and hearing loss [2, 5].

Cx hemichannels (HCs) can selectively open to exchange inorganic cations and anions, metabolites, and signaling molecules between the cytoplasm and the extracellular milieu [6]. Leaky or hyperactive HCs may cause disruption of transcellular ion gradients, loss of critical metabolites, influx of Ca^{2+}, and release of ATP [3, 7]. The latter plays a prominent role in the initiation of inflammation, and the role of HCs in genetically determined inflammatory and hyperproliferative skin diseases is well documented [8]. Moreover, disruption of Ca^{2+} homeostasis and increased ATP release impact on epidermal renewal and barrier formation in the skin [9, 10], as well as on organ development and hearing acquisition in the inner ear [11–13].

The prognosis for KID syndrome is variable, with patients suffering from blindness, deafness, and disfigurement. Skin abnormalities include unusually thick skin on the palms of the hands and soles of the feet (palmoplantar keratoderma); thick, red patches of skin (erythrokeratoderma); and dry, scaly skin (ichthyosis). Ichthyotic skin has decreased barrier function and a propensity for transepidermal water loss and infections. KID syndrome can also lead to an aggressive form of squamous cell carcinoma. Eye problems are caused by keratitis (corneal inflammation) which can lead to pain, sensitivity to light, extra blood vessel growth, scarring, and eventual vision loss or blindness [1]. The G45E and A88V mutations of Cx26 cause 100% lethality in infancy [14, 15].

Mice with inducible expression of the Cx26-G45E mutation in keratinocytes replicate the epidermal pathology of KID syndrome [16], allowing epidermal pathology to be evaluated using an in vivo imaging system [17, 18]. In this chapter, we provide detailed instructions to automatically detect and quantify epidermal lesions in Cx26-G45E mice based on ratiometric reflectance imaging (Fig. 1). The system and procedures we developed are useful to determine the efficacy of treatment for this devastating condition, as demonstrated for the abEC1.1 monoclonal antibody [18]. When combined with advanced intravital multiphoton microscopy [19], this approach can provide invaluable insight into the molecular underpinnings of skin pathophysiology and the effects of treatment [20].

2 Materials

2.1 Animals

1. Cx26-G45E double transgenic mice (MGI:5635532).

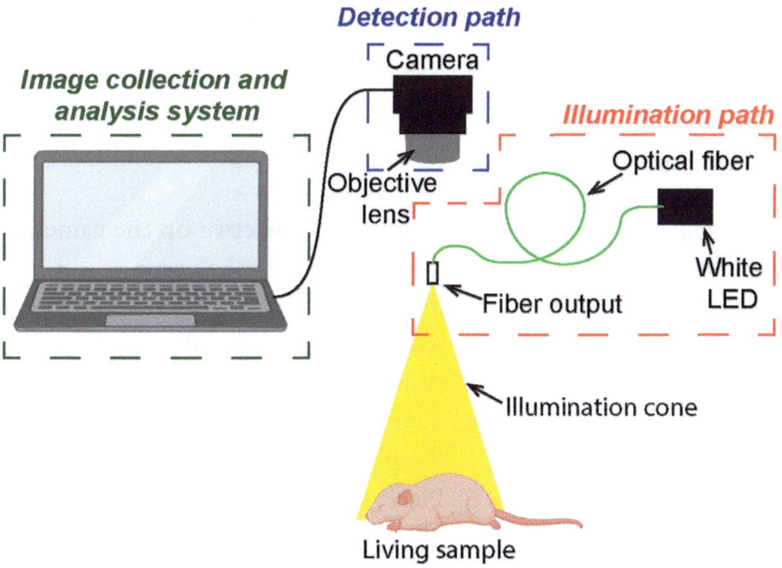

Fig. 1 Simplified scheme of the imaging and analysis system based on ratiometric reflectance imaging

2.2 Optical Components

2. Optical breadboard or optical table.
3. Warm white light emitting diode (e.g., MWWHF2, Thorlabs Inc.).
4. Multimode fiber for LED coupling (e.g., Ø1000 μm, M35L02, Thorlabs).
5. Objective lens with adjustable focus system (e.g., TV Lens, F1.2 8 mm, Ernitec).
6. Digital camera with a cMOS color sensor (e.g., DCC1645C, Thorlabs).
7. White or gray reflective target.
8. Fiber output L-shaped holder with adjustable height, position, and orientation.
9. Camera L-shaped holder with adjustable height.
10. Phantom sample of about same sizes of a mouse.

2.3 Anesthesia and Homeothermic System

1. Stand-alone gaseous anesthesia system (e.g., 901,806, V-1 Tabletop Laboratory Animal Anesthesia System, VetEquipe Inc.).
2. Feedback-controlled heating system (e.g., ThermoStar Homeothermic Blanket-Monitoring System).

2.4 Image Collection and Analysis System

1. Camera control software for white balancing and exposure time adjustment (e.g., 64-Bit ThorCam Software for Windows).

2. ImageJ/Fiji Software (ver. 1.53 or later).
3. Fiji script for image analysis (*see* Subheading 3.4).

3 Methods

3.1 Optical Alignment and Imaging Adjustments

1. Mount the objective on the camera (detection system).
2. Mount the detection system and the fiber output on L-shaped holders.
3. Assemble the optical components and anesthesia system as shown in Fig. 2 (*see* **Note 1**).
4. Switch on the LED light and place the mouse phantom under the illumination cone in front of the anesthesia system.
5. Place the white or grey reflective target close to the mouse phantom (*see* **Note 2**).
6. Adjust the fiber output elevation over the mouse phantom to generate a cone of illumination wide enough to illuminate uniformly both the phantom and the reflective target.
7. Adjust the detection system elevation so that both the mouse phantom and the reflective target are encompassed within the camera field of view.
8. Optimize detection system elevation to focus sharply on the mouse phantom (*see* **Note 3**).
9. To avoid image saturation, adjust the LED drive current while imaging the reflective target.
10. Perform camera white balancing calibration using a white target.
11. Manually adjust the camera RGB calibration to easily detect red skin patches on the mouse back (*see* **Note 4**).

3.2 Animal Preparation

1. Remove the mouse from its cage and assess its overall well-being. Avoid this procedure on animals with breathing problems or clearly unhealthy (*see* **Notes 5**–7).
2. Preanesthetize the mouse using a precision vaporizer system connected to an oxygen tank. Place each mouse inside the induction chamber and seal/close the top for ~1 min at 4% isoflurane in 100% oxygen.
3. Switch system flow and transfer mice to the anesthesia mask/nose cone under the illumination cone and adjust gas flow with vaporizer to 2% isoflurane (*see* **Note 8**).

3.3 Assessment of Lesions Detection Capabilities

1. Switch on the warm white LED light 1 h before image acquisition to ensure stability and minimize light intensity fluctuation.

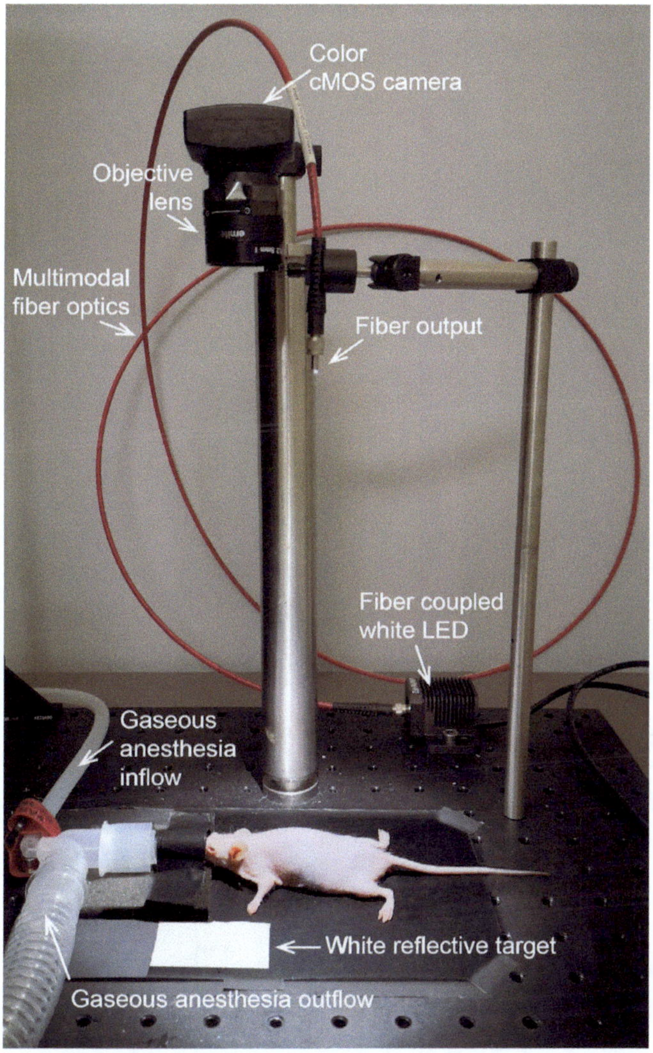

Fig. 2 Photograph of the system components

2. Prepare the mouse as described in Subheading 3.2. To assess the system lesions detection capabilities, perform imaging on both untreated and treated Cx26G45E mice (*see* **Note 9** and Fig. 3).
3. By using the camera software, look at the computer screen and adjust the mouse position to achieve an even illumination of the entire mouse back skin. Pay attention that all four paws are visible within the field of view and none of them is placed under the animal.
4. For each mouse, acquire a set of 40 images. Acquisition rate must be optimized based on the animal breathing rate. We usually acquire at a frequency of 0.9 frames/s.

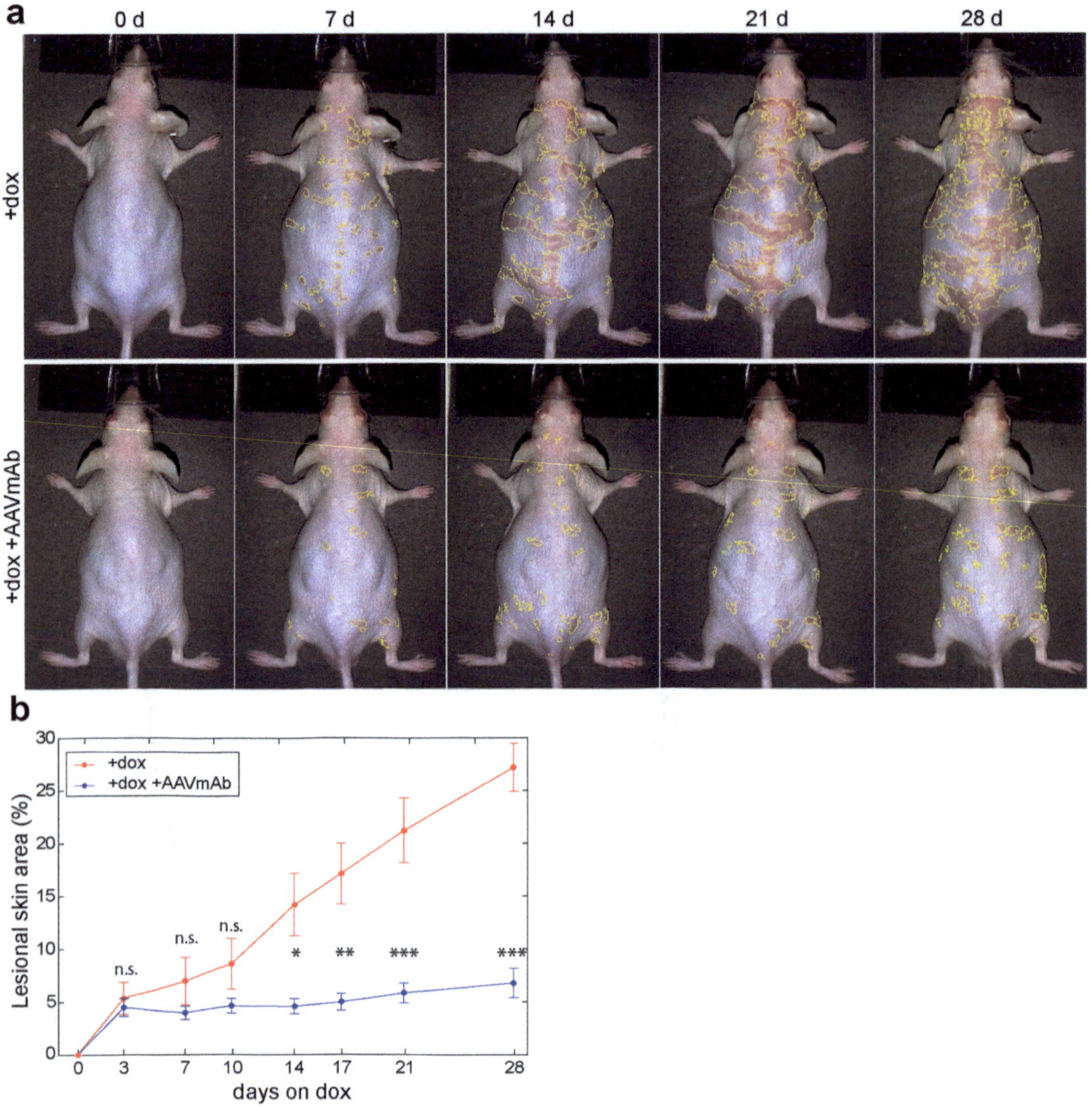

Fig. 3 The system automatically detects and quantifies the visible manifestation of epidermal lesions. (**a**) Representative examples of epidermal lesions automatically detected by ratiometric reflectance image analysis at indicated time points after dox induction. Top: untreated dox-induced Cx26-G45E transgenic mouse (+dox) showing typical KID pathology worsening over time. Skin lesions are shown contoured by yellow lines automatically drawn by the analysis algorithm. Bottom: dox-induced Cx26-G45E transgenic mouse treated with AAVmAb (+dox + AAVmAb); treatment limited the skin area affected by lesions. (**b**) Quantification of the lesional skin area vs. days on dox. Data are shown as mean $\pm$ standard error of the mean (s.e.m.) for untreated ($n = 6$) and treated ($n = 6$) animals; p-values (P) were computed with Mann–Whitney U statistical test. *, $P < 0.05$; **, $P < 0.01$; ***, $P < 0.005$; $P > 0.05$ was considered not significant (n.s)

5. The lesion analysis is performed on an image obtained by averaging. Discard images with evident mouse breathing movements. To achieve the required signal-to-noise ratio, at least 20 breathing artifacts-free images are required.

6. Switch off the anesthesia and keep the mouse on the heating pad during recovery. Fast recovery is crucial for repeated experiments. If animals require more than 3 min for a full recovery, reduce anesthesia dosage.
7. Place animal back into cage and monitor for 30 min.
8. Repeat this procedure for each mouse.

3.4 Image Analysis

1. Download the Fiji script used for epidermal lesions quantification available from the GitHub repository at https://github.com/chiaraperes/AAVmAb_effect/blob/main/LesionExtentEvaluation.ijm
2. Modify the image name and parameters as requested in the script.
3. Run the script on your set of images (*see* **Note 10** and Fig. 4). The Fiji script algorithm returns the list of the identified lesion and their yellow contour on the analyzed image (Fig. 4, **Step 9**).
4. If present, manually discard the contours outlining ears, paws, and tail.
5. Finally, run the "*Multi Measure*" plugin in ROI manager, *roiManager("Multi Measure")*, to measure the extension of ROIs encompassed by all the remaining contours. This crucial step returns the list of the lesion sizes in pixels unit.

4 Notes

1. To maximize image quality, both illumination and detection systems should be aligned along the same optical axis. In our architecture, the fiber output is slightly offset laterally.
2. A reflective target is used to verify uniformity and stability of image acquisition. It can also be used to rescale red or blue channel intensity of the acquired mouse images in case of significant light intensity fluctuation. The image analysis script contains a section to evaluate RGB channel light intensity reflected by the target.
3. Adjust the detection system height for coarse focusing. Use the objective fine focus system to optimize focusing before starting data acquisition.
4. This step strongly depends on the illumination light and detection system. To be sure to easily detect skin lesions, put a wild-type mouse under the illumination cone and adjust red and blue channel gain to have an image as close as possible to true color. Then slightly reduce the red channel gain obtaining a

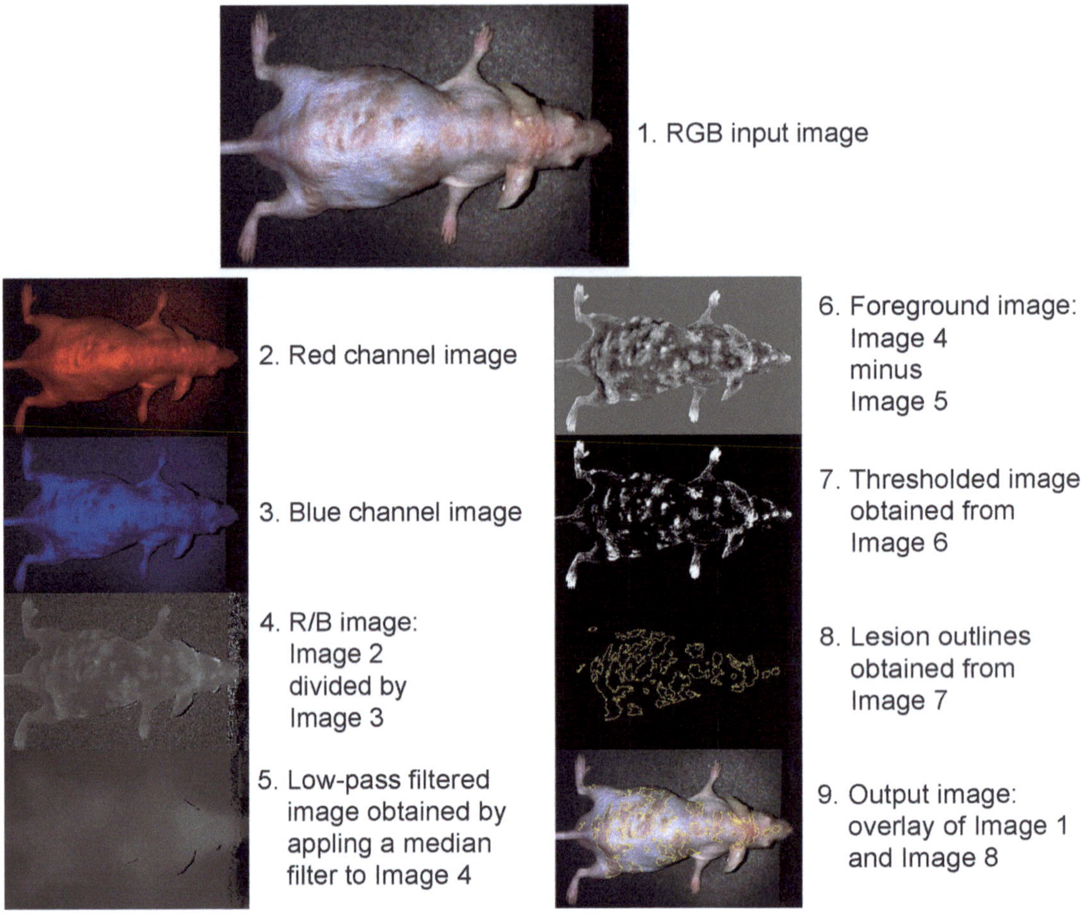

Fig. 4 Main steps of image analysis used to automatically estimate lesional area (projected on the image plane)

bluish image of the mouse: it eases detection of red skin patches.

5. All animal procedures were agreed upon, reviewed and approved by local animal welfare oversight bodies and were performed with the approval and direct supervision of the CNR-IBBC/Infrafrontier—Animal Welfare and Ethical Review Body (AWERB), in accordance with general guidelines regarding animal experimentation, approved by the Italian Ministry of Health, in compliance with the Legislative Decree 26/2014 (ref. Project license 603/2018-PR and 632/2019-PR), transposing the 2010/63/EU Directive on protection of animals used in research) and in adherence to the NIH Guide for the Care and Use of Laboratory Animals and recommendations from both ARRIVE and PREPARE guidelines [21, 22].
6. When a doxycycline (dox) containing diet is administered (625 mg/kg, Mucedola, Settimo Milanese, Italy), these mice

express the mutant Cx26-G45E and EGFP as independent proteins in epidermal keratinocytes. After few days from dox induction, transgenic Cx26-G45E mice develop skin abnormalities and epidermal pathology worsens progressively when mice are maintained on dox diet for longer periods. Mice expressing Cx26-G45E in the skin have reduced viability, hyperkeratosis, scaling, skin folds, and hair loss [16], together with thick, red patches of skin (erythrokeratoderma). TheCx26-G45E mouse model is characterized by a large phenotypic viability, such that only about 3% of the bred animals can be used for experiments, as reported previously [17]. Accordingly, more than 400 mice had to be screened over a 3-year period to achieve the cohorts with similar disease progression used for the present experimental studies. During the screening process aimed at isolating animals with similar disease progression, we noted that female mice showed a significantly milder epidermal expression of pathological lesions. This is probably related to our mating strategy. For this reason, only adult male mice, aged between 5 and 17 weeks, were used for experiments [23].

7. To avoid motion artifacts during image collection, to simplify animal positioning and reduce animal discomfort, we used anesthetized animals. Since the complete procedure (animal positioning and image acquisition) requires 10 min maximum and it is meant to be repeated more than once per week, we used gaseous anesthesia due to ease of use and faster animal recovery. To maximize animal comfort and speed up recovery, our system is also endowed with a feedback-controlled heating pad for homeothermic regulation of mouse body temperature.
8. The animal is sufficiently anesthetized when the breathing rate slows down and becomes constant. It is not required to reach a surgical plane of anesthesia (assessed by firm toe pinch).
9. The ratiometric reflectance imaging system was used to evaluate the effect of in vivo treatment on Cx26-G45E mice with an AAV encoding for the expression of abEC1.1, an antagonist monoclonal antibody targeting the extracellular domain of Cx HCs that has been a previously characterized and validated both in vitro and in vivo [24–26]. We tested if the optical architecture and analysis method can be used to automatically detect and quantify the reduction of visible manifestation of epidermal pathology in Cx26-G45E mice treated with the AAV (Fig. 3).
10. The script separates the three color channels of each image, thus generating three corresponding grayscale 8-bit images (each one with pixel intensity values comprised between 0 and 255). For automated lesion identification, the script

evaluates the pixel-by-pixel ratio of the R channel divided by the B channel (R/B intensity image). A 60 × 60 pixels median filter is then applied to the R/B image, thus generating a heavily low pass (lp) filtered version of it (R/B_lp). This image is used as a background image that is subtracted from the R/B image. Next, in the script, we apply an intensity threshold to the resulting high-pass (hp) filtered image (R/B_hp = R/B − R/B_lp) and we feed the result to the "*Analyze particle*" plugin to determine lesion contours (*see* Fig. 4).

Acknowledgments

This work was funded by grants awarded to Fabio Mammano by Fondazione Telethon (GGP19148) and the Italian Ministry of Education, Universities and Research (PRIN 20175C22WM).

References

1. Sybert VP (2017) Genetic skin disorders. Oxford monographs on medical genetics. Oxford University Press
2. Lilly E, Sellitto C, Milstone LM et al (2016) Connexin channels in congenital skin disorders. Semin Cell Dev Biol 50:4–12. https://doi.org/10.1016/j.semcdb.2015.11.018
3. Garcia IE, Bosen F, Mujica P et al (2016) From hyperactive Connexin26 Hemichannels to impairments in epidermal calcium gradient and permeability barrier in the keratitis-ichthyosis-deafness syndrome. J Invest Dermatol 136(3):574–583. https://doi.org/10.1016/j.jid.2015.11.017
4. Laird DW, Lampe PD (2022) Cellular mechanisms of connexin-based inherited diseases. Trends Cell Biol 32(1):58–69. https://doi.org/10.1016/j.tcb.2021.07.007
5. Lilly E, Bunick CG, Maley AM et al (2019) More than keratitis, ichthyosis, and deafness: multisystem effects of lethal GJB2 mutations. J Am Acad Dermatol 80(3):617–625. https://doi.org/10.1016/j.jaad.2018.09.042
6. Lucaciu SA, Leighton SE, Hauser A et al (2023) Diversity in Connexin biology. J Biol Chem 299:105263. https://doi.org/10.1016/j.jbc.2023.105263
7. Saez JC, Schalper KA, Retamal MA et al (2010) Cell membrane permeabilization via connexin hemichannels in living and dying cells. Exp Cell Res 316(15):2377–2389. https://doi.org/10.1016/j.yexcr.2010.05.026
8. Levit NA, White TW (2015) Connexin hemichannels influence genetically determined inflammatory and hyperproliferative skin diseases. Pharmacol Res 99:337–343. https://doi.org/10.1016/j.phrs.2015.07.015
9. Burnstock G, Knight GE, Greig AV (2012) Purinergic signaling in healthy and diseased skin. J Invest Dermatol 132(3 Pt 1):526–546. https://doi.org/10.1038/jid.2011.344
10. Essenfelder GM, Bruzzone R, Lamartine J et al (2004) Connexin30 mutations responsible for hidrotic ectodermal dysplasia cause abnormal hemichannel activity. Hum Mol Genet 13(16):1703–1714. https://doi.org/10.1093/hmg/ddh191
11. Mammano F, Bortolozzi M (2018) Ca(2+) signaling, apoptosis and autophagy in the developing cochlea: milestones to hearing acquisition. Cell Calcium 70:117–126. https://doi.org/10.1016/j.ceca.2017.05.006
12. Mazzarda F, D'Elia A, Massari R et al (2020) Organ-on-chip model shows that ATP release through connexin hemichannels drives spontaneous Ca(2+) signaling in non-sensory cells of the greater epithelial ridge in the developing cochlea. Lab Chip 20(16):3011–3023. https://doi.org/10.1039/d0lc00427h
13. Sun L, Gao D, Chen J et al (2022) Failure of hearing acquisition in Mice with reduced expression of Connexin 26 correlates with the abnormal phasing of apoptosis relative to autophagy and defective ATP-dependent Ca (2+) signaling in Kolliker's organ. Front Cell

Neurosci 16:816079. https://doi.org/10.3389/fncel.2022.816079

14. Janecke AR, Hennies HC, Gunther B et al (2005) GJB2 mutations in keratitis-ichthyosis-deafness syndrome including its fatal form. Am J Med Genet A 133A(2):128–131. https://doi.org/10.1002/ajmg.a.30515
15. Haruna K, Suga Y, Oizumi A et al (2010) Severe form of keratitis-ichthyosis-deafness (KID) syndrome associated with septic complications. J Dermatol 37(7):680–682. https://doi.org/10.1111/j.1346-8138.2010.00839.x
16. Mese G, Sellitto C, Li L et al (2011) The Cx26-G45E mutation displays increased hemichannel activity in a mouse model of the lethal form of keratitis-ichthyosis-deafness syndrome. Mol Biol Cell 22(24):4776–4786. https://doi.org/10.1091/mbc.E11-09-0778
17. Sellitto C, Li L, White TW (2021) Connexin hemichannel inhibition ameliorates epidermal pathology in a mouse model of keratitis ichthyosis deafness syndrome. Sci Rep 11(1):24118. https://doi.org/10.1038/s41598-021-03627-8
18. Peres C, Sellitto C, Nardin C et al (2023) Antibody gene transfer treatment drastically improves epidermal pathology in a keratitis ichthyosis deafness syndrome model using male mice. EBioMedicine 89:104453. https://doi.org/10.1016/j.ebiom.2023.104453
19. Peres C, Nardin C, Yang G et al (2022) Commercially derived versatile optical architecture for two-photon STED, wavelength mixing and label-free microscopy. Biomed Opt Express 13(3):1410. https://doi.org/10.1364/boe.444525
20. Donati V, Peres C, Nardin C et al (2022) Calcium signaling in the photodamaged skin: in vivo experiments and mathematical modeling. Function 3(1):zqab064. https://doi.org/10.1093/function/zqab064
21. Kilkenny C, Browne WJ, Cuthill IC et al (2010) Improving bioscience research reporting: the ARRIVE guidelines for reporting animal research. PLoS Biol 8(6):e1000412. https://doi.org/10.1371/journal.pbio.1000412
22. Smith AJ, Clutton RE, Lilley E et al (2018) PREPARE: guidelines for planning animal research and testing. Lab Anim 52(2):135–141. https://doi.org/10.1177/0023677217724823
23. Xu H, Gu S, Riquelme MA et al (2015) Connexin 43 channels are essential for normal bone structure and osteocyte viability. J Bone Miner Res 30(3):436–448. https://doi.org/10.1002/jbmr.2374
24. Xu L, Carrer A, Zonta F et al (2017) Design and characterization of a human monoclonal antibody that modulates mutant Connexin 26 Hemichannels implicated in deafness and skin disorders. Front Mol Neurosci 10:298. https://doi.org/10.3389/fnmol.2017.00298
25. Ziraldo G, Buratto D, Kuang Y et al (2019) A human-derived monoclonal antibody targeting extracellular Connexin domain selectively modulates Hemichannel function. Front Physiol 10:392. https://doi.org/10.3389/fphys.2019.00392
26. Nardin C, Tettey-Matey A, Donati V et al (2022) A quantitative assay for Ca(2+) uptake through normal and pathological Hemichannels. Int J Mol Sci 23(13). https://doi.org/10.3390/ijms23137337

Chapter 14

Measuring Connexin Hemichannel Opening in Response to an InsP3-Mediated Cytosolic Ca^{2+} Increase

Erva Bayraktar and Mario Bortolozzi

Abstract

The opening of connexin hemichannels (HCs) expressed at the plasma membrane of mammalian cells is regulated by a number of physiological parameters, including extracellular and intracellular Ca^{2+} ions. Submicromolar variations of the cytosolic Ca^{2+} concentration ($[Ca^{2+}]_c$) are per se sufficient to trigger extracellular bursts of messenger molecules through connexin HCs, thus mediating paracrine signaling. In this chapter, we present a quantitative method to measure the opening dynamics of connexin HCs expressed in a single HeLa cell upon stimulation by a canonical InsP3-mediated $[Ca^{2+}]_c$ transient. The protocol relies on a combination of Ca^{2+} imaging and patch-clamp techniques. The insights gained from our method are expected to make a significant contribution to understanding the structure–function relationship of connexin HCs. The protocol is also suitable to screen candidate therapeutic compounds to treat connexin-related diseases linked to HC dysfunction.

Key words Connexin 32, Hemichannel conductance, HeLa cells, Gating by cytosolic Ca^{2+}, Patch-clamp

1 Introduction

Connexins are a family of tetraspan membrane proteins that form hexameric structures, called connexons or hemichannels (HCs) [1]. Two HCs belonging to adjacent cells can dock end-to-end to form a gap junction (GJ) channel [2], which serves as conduit for the transfer of ions and messenger molecules up to 1 kDa size [3]. The undocked HC is typically closed at rest, since its large and relatively unselective pore mediates sustained fluxes of ions and messenger molecules, such as ATP [4], cAMP [5], glutamate [6, 7], NAD^+ [8, 9], glutathione [10], D-serine [11, 12], and prostaglandins [13, 14]. These molecules are involved in autocrine/paracrine cell communication regulating development, homeostasis, function, and regeneration of several organs.

The opening of connexin HCs is mainly regulated by the synergistic action of membrane potential and Ca^{2+} ions [15, 16]

Fabio Mammano and Mauricio Retamal (eds.), *Connexin Hemichannels: Methods and Protocols*, Methods in Molecular Biology, vol. 2801, https://doi.org/10.1007/978-1-0716-3842-2_14,

(both extracellular Ca^{2+} [17, 18] and cytosolic Ca^{2+} [19–21]). Typically, ATP released by a cell through connexin HCs diffuses extracellularly and binds to P2Y receptors of neighboring cells activating a canonical inositol 1,4,5-trisphosphate (InsP3)-mediated transduction cascade [4]. The opening of InsP3 receptors expressed at the endoplasmic reticulum (ER) and the Golgi apparatus allows the release of Ca^{2+} that determines the raise of the cytosolic Ca^{2+} concentration ($[Ca^{2+}]_c$). In turn, this $[Ca^{2+}]_c$ elevation can open new HCs, propagating the Ca^{2+} signalling. HCs display a bell-shaped open probability on $[Ca^{2+}]_c$, that is, nearly zero at resting concentration and maximal at 500 nM [20, 21].

Here, we present a method we have developed [19] to electrophysiologically measure the opening and closure of connexin HCs in response to an InsP3-mediated $[Ca^{2+}]_c$ elevation. The protocol is performed in single HeLa cells transfected with the 32 kDa human connexin 32 (Cx32), a connexin isoform whose mutations are the leading cause of the X-linked dominant form of Charcot–Marie–Tooth disease (CMTX1) [22]. In the peripheral nervous system, Cx32 expressed in myelinating Schwann cells was proposed to play a key role in the homeostasis and integrity of the myelinated axons through the formation of minute intracellular GJs along the myelin sheath [23, 24]. Recent evidence suggests that permeability or gating alterations of Cx32 HCs may be an alternative mechanism for the CMT1X pathogenesis. In fact, some CMTX1 mutations retain the ability to ensure normal GJ-mediated cell–cell coupling in vitro while exhibiting HC defects [19, 25]. These findings support the hypothesis that alteration of Cx32 HC opening and ATP release could underlie CMTX1. Indeed, an ATP-mediated paracrine signaling in Schwann cells is critical for the myelination process [26–29].

In our protocol, Ca^{2+} release from the intracellular stores of a single HeLa cell expressing wild-type (WT) or mutant Cx32 HCs was stimulated by an ATP (10 μM) or histamine (250 μM) puff. The $[Ca^{2+}]_c$ transient was optically measured by Fura-red or Fura-2 Ca^{2+} dyes, respectively, while the rapid increase in membrane conductance (G_m) due to Cx32 HC opening was measured by patch-clamp recording in voltage-clamp configuration (Fig. 1). Immediately before the ATP/histamine puff, the average membrane resistance (R_m) of the patched cell was >2 GΩ (>4 GΩ for untransfected cells), thus corresponding to a nearly complete closure of Cx32 HCs. During the puff, the G_m variation (ΔG_m) lasted a few seconds and had exponential behavior with the rising time τ (i.e., HC opening) significantly shorter than the decay time (i.e., HC closure) for both ATP (0.86 s vs. 1.16 s) and histamine (1.49 s vs. 2.43 s) stimulations. The experiments with the nonspecific connexin HC blocker carbenoxolone (CBX, 100 μM) and the selective Cx32 HC blocker GAP24 (a mimetic peptide with

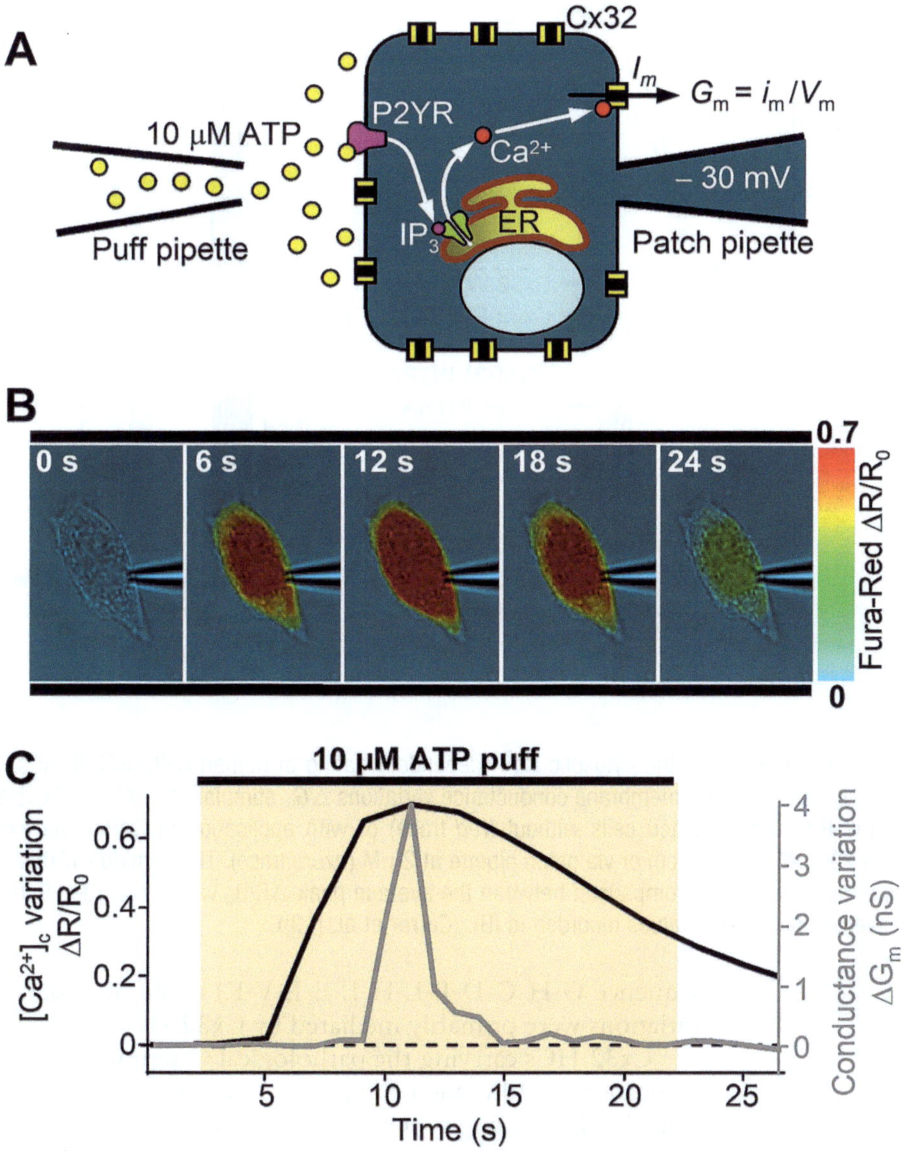

Fig. 1 Measuring opening and closure of Cx32 HCs expressed in HeLa cells in response to an InsP3-mediated $[Ca^{2+}]_c$ transient. (**a**) Scheme of the patch-clamp experiment to study the HC gating by $[Ca^{2+}]_c$ in a single Cx32-WT transfected HeLa cell. An InsP3-dependent $[Ca^{2+}]_c$ transient is stimulated by an extracellular puff containing 10 μM ATP. Cx32 HC opening and closure were monitored in terms of ΔG_m computed by the periodic application (at 1 Hz) of a +10 mV voltage step lasting 100 ms. The contribution to ΔG_m by other Ca^{2+}-activated channels was kept negligible by the addition of specific blockers to the extracellular solution. (**b**) Representative frame sequence of an experiment using the Fura-Red $[Ca^{2+}]_c$ dye. (**c**) Time course of the same experiment in (b) in terms of Fura-red $\Delta R/R_0$ signal and membrane conductance variation ΔG_m during the opening and closure of Cx32-WT HCs. (Carrer et al. [19])

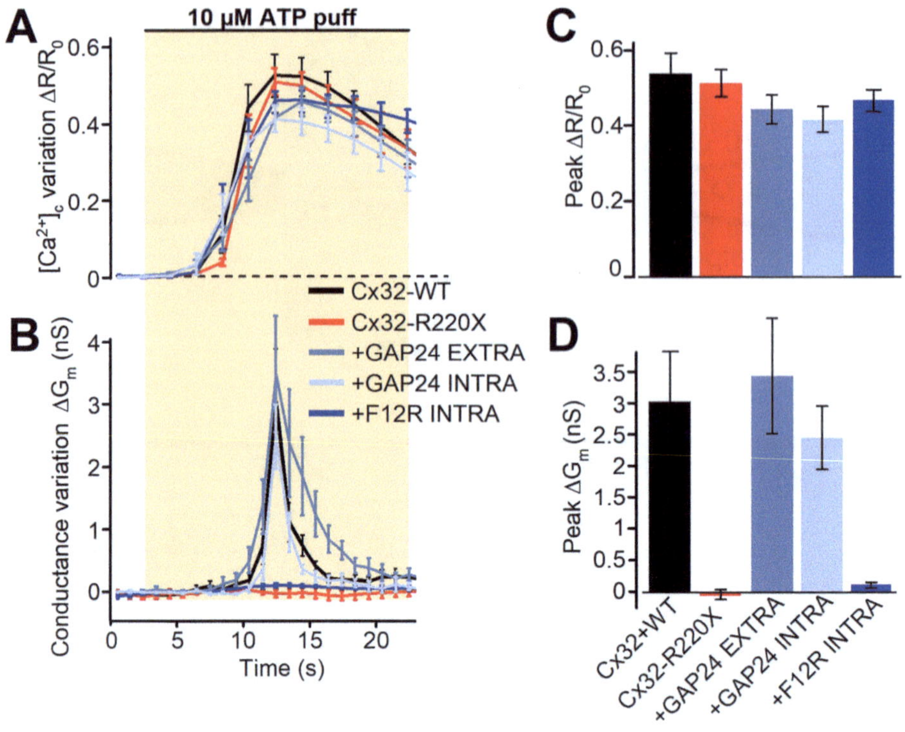

Fig. 2 GAP24 peptide restores the cytosolic Ca^{2+}-dependent gating of human Cx32-R220X HCs. (**a** and **b**) Average $[Ca^{2+}]_c$ transients and membrane conductance variations ΔG_m stimulated by ATP in Cx32-WT (*black* trace) or Cx32-R220X transfected cells without (*red* trace) or with application of GAP24 via extracellular incubation at 200 μM (*violet* trace) or via patch pipette at 2 μM (*azure* trace). The peptide F12R is a negative control (*blue* trace). (**c** and **d**) Comparison between the average peak $\Delta R/R_0$ values recorded in (A) together with their average peak ΔG_m values recorded in (B). (Carrer et al. [19])

sequence G-H-G-D-P-L-H-L-E-E-V-K) confirmed that these G_m variations were primarily mediated by Cx32 HCs.

Cx32 HCs carrying the pathological C-terminus (CT) domain truncation R220X failed to open in response to $[Ca^{2+}]_c$ elevations (Fig. 2). Interestingly, the gating function of Cx32-R220X HCs was completely restored by both the intracellular and extracellular application of GAP24, suggesting that the CT domain of Cx32 is not directly involved in the gating mechanism but acts as a modulator. Molecular dynamics simulations indicated that this paradoxical recovery effect mediated by GAP24 in R220X HCs could be mediated by direct interaction of the peptide with the HC cytoplasmic loop [19].

Overall, our protocol proved useful not only in studying the structure–function relationship of connexin HCs but also in screening candidate therapeutic compounds to treat connexin-related diseases linked to HC dysfunction.

2 Materials

2.1 Cell Culture and Cx32 Protein Expression in HeLa Cells

1. HeLa cells (DH clone, Sigma-Aldrich).
2. Dulbecco's modified eagle's medium (DMEM, Gibco, Thermo Fisher Scientific) supplemented with 10% fetal bovine serum (FBS) (Gibco) and 1% penicillin- streptomycin (5000 U/mL, Gibco) for cell culture.
3. OptiMEM reduced serum medium (Gibco).
4. Lipofectamine 3000 transfection kit (Thermo Fisher Scientific).
5. Cell culture 24-well dishes and 12 mm diameter glass coverslips.
6. A Cx32-IRES-EGFP expression vector carrying the coding region of human Cx32-WT (plasmid #78150, Addgene).
7. Humidified incubator (37 °C, 90% humidity, 5% CO_2).
8. BSL1 laminar flow.

2.2 Setting Up Imaging and Electrophysiological Experiments

1. An experimental chamber mounted on the stage of an upright wide-field fluorescence microscope (BX51, Olympus) with an infinity-corrected water immersion objective (40×, 0.8 NA, Olympus) (*see* **Note 1**).
2. A gravimetric filling system or a peristaltic pump to superfuse the cells with experimental solutions.
3. An amplifier (Axon Instruments, MultiClamp 700B) with a dual headstage configuration capable of carrying out simultaneous measurements.
4. A digitizer (Axon Instruments, Digidata® 1550A Low-Noise Data Acquisition System).
5. Borosilicate glass capillaries (GC150–15 and G85150T-4, Harvard Apparatus).
6. A single stage glass microelectrode puller (PP-830, Narishige).
7. A Monochromator (Polychrome V, Till Photonics) and/or collimated LEDS (Thorlabs) along with appropriate excitation and emission filters (Chroma).
8. Software for data acquisition and analysis (*see* **Note 2**).

2.3 Simultaneous Recording of Cytosolic Ca^{2+} Changes and Cell Membrane Conductance Variations (upon ATP or Histamine Stimuli) (See Note 3)

1. Extracellular solution: 140 mM NaCl, 5 mM KCl, 10 mM HEPES, 0.5 mM $CaCl_2$, 2 mM sodium pyruvate and 5 mM glucose, 4 mM tetraethylammonium chloride (TEA-Cl), 4 mM CsCl, 2 mM 4-aminopyridine (4-AP), 2 μM TRAM-34, 400 nM UCL-1684 (pH 7.4, 328 mOsm) (*see* **Notes 4** and **5**).
2. Intracellular solution: 130 mM potassium aspartate (KAsp), 10 mM NaCl, 10 mM KCl, 10 mM HEPES, 1 mM $MgCl_2$, 50 μM BAPTA (pH 7.2, 313 mOsm) (*see* **Note 6**).

3. Puff solution: 10 μM ATP or 250 μM histamine diluted in extracellular solution.
4. For calcium imaging: 10 μM Fura-2 acetoxymethyl esters (Molecular Probes, Thermo Fisher Scientific), Pluronic F-127 (0.1%, w/v) and 250 μM sulfinpyrazone in DMEM.
5. For control experiments: nonspecific connexin blocker carbenoxolone (CBX) or selective Cx32 HC blocker GAP24 (amino acid sequence: G-H-G-D-P-L-H-L-E-E-V-K, Alpha Diagnostic International) (*see* **Note 7**).

3 Methods

3.1 Cell Culture and Cx32 Protein Expression in HeLa Cells

1. Plate HeLa cells (10,000 cells) onto 12 mm diameter glass coverslips in a 24-well dish in DMEM supplemented with 10% FBS and 1% penicillin–streptomycin (5000 U/mL) at 37 °C in a humidified incubator with 5% CO_2.
2. One day after plating the cells, transfect them with a pIRES2-EGFP expression vector carrying the coding region of human Cx32 (RefSeqGene: NG_008357.1) using Lipofectamine 3000.

3.1.1 Lipofectamine 3000 Transfection Protocol (for 1 Coverslip in a 24 Multiwell, Well Area 1.9 cm²)

1. Equilibrate Lipofectamine 3000, P3000 Reagent, 50 μL of OptiMEM, and plasmid DNA to room temperature.
2. Substitute the medium inside a well with 450 μL of fresh DMEM supplemented with 10% FBS (*see* **Note 8**).
3. Divide the OptiMEM into two 25 μL aliquots. In the first aliquot, slowly add 0.8 μL of Lipofectamine 3000 and mix gently.
4. In the second aliquot, add 0.4 μg of plasmid DNA and 0.8 μL of P3000 Reagent and mix properly (vortexing at low speed).
5. After 5 min incubation, add the second mixture into the one with Lipofectamine 3000 and mix slowly.
6. Let the DNA-lipid mixture rest for (at least) 5′.
7. Add the DNA-lipid complex into the top of the cells drop by drop. Return the plate to the incubator without shaking.
8. Incubate the cells for a minimum of 2.5–4 h at 37 °C.
9. Change the contents of the well with 500 μL of complete culture medium.

3.2 Setting Up Imaging and Electrophysiological Experiments (See Note 9)

1. Live imaging and electrophysiological experiments are performed 24 h after transfection.
2. Prior to the experiment, cells are loaded with 10 μM Fura-2 acetoxymethyl in DMEM with Pluronic F-127 (0.1%, w/v) and 250 μM sulfinpyrazone via a 30′ incubation at 37 °C.

3. Once loaded, a glass coverslip with loaded transfected cells is transferred to the experimental chamber (described above) and superfused with extracellular solution (at the speed of 1 mL/min) at room temperature (22–24 °C).
4. A single stage puller is used to obtain patch pipettes from borosilicate glass capillaries with a final DC resistance of patch electrodes ranging from 4 to 9 MOhm (depending on the glass type and the solutions utilized in the experiment).
5. Cytosolic GFP fluorescence is excited at 480 nm to identify isolated transfected cells to perform subsequent imaging/electrophysiology experiments.

3.3 Simultaneous Recording of Cytosolic Ca^{2+} Changes and Cell Membrane Conductance Variations (upon ATP or Histamine Stimuli)

1. Allow at least a 20′ de-esterification postloading with the calcium indicator in the extracellular solution, after which, the selected GFP-positive cell is contacted by a patch pipette filled with the intracellular solution and the whole-cell configuration is achieved while monitoring Fura fluorescence to check for unwanted $[Ca^{2+}]_c$ transients stimulated by the mechanical stretch of the cell.
2. Within a minute after achieving the whole-cell configuration and maintaining the cell at a holding potential of −30 mV (approximately its resting potential), a puff with 10 μM ATP or 250 μM histamine, diluted in extracellular solution, is gently applied for 20 s using a pipette positioned close (50 μm) to the selected cell to elicit a physiological $[Ca^{2+}]_c$ increase.
3. The Cx32 HC opening is monitored in terms of ΔG_m computed by the periodic application (at 1 Hz) of +10 mV voltage step lasting 100 ms (*see* **Note 10**).
4. The Ca^{2+} dynamics is monitored in terms of Fura values $\Delta R/R_0 = (R(t)\text{-}R_0)/R_0$, averaged in a region of pixels correspondent to the patched cell, where $R(t)$ is Fura ratio at time t and R_0 is the prestimulus ratio (*see* **Note 11**).
5. For control experiments: the cells are incubated with the nonspecific connexin blocker CBX in the extracellular solution for 30′ at 100 μM concentration, followed by a 5′ washout before contacting the selected cell by the patch pipette. Alternatively, the cells can be incubated with the selective Cx32 HC blocker GAP24 in the extracellular solution for 30′ at 200 μM concentration, followed by a 5′ washout before contacting the selected cell by patch pipette.

4 Notes

1. The microscope was equipped with a sCMOS camera (PCO Edge 5.5) to approach the cell by the patch pipette using brightfield illumination and then to acquire the Ca^{2+} dye fluorescence.

2. Electrophysiological data were acquired by pClamp software (version 10.4, Molecular Device). Electrophysiological and imaging data were analyzed with a software developed in our laboratory using Matlab tools (The MathWorks, Inc.).
3. All chemicals were purchased from Sigma-Aldrich unless otherwise specified. Solubility varies depending on the specific salt form, so please refer to the product specification sheet for details.
4. To favor Cx32 HC opening during the $[Ca^{2+}]_c$ increase, the extracellular Ca^{2+} concentration was maintained at 500 μM.
5. Any contribution to ΔG_m by other Ca^{2+}-activated channels is kept negligible by the addition of specific blockers in the extracellular solution.
6. 50 μM Fura-2 pentapotassium salt is utilized instead of 50 μM BAPTA in the experiments with the histamine puff. After preparing the intracellular solution, filter it through a 0.22 μm pore size membrane (Millipore).
7. Lyophilized GAP24 powder may be stored at −20 °C and remain stable for up to 12 months. Reconstitute the powder with ddH_2O and aliquot it to avoid repeated freezing and thawing, then store it at −20 °C. The reconstituted product is stable for up to 6 months at −20 °C.
8. To wash out the antibiotics which can be toxic in the presence of lipofectamine.
9. This design allows us the study of the opening/closure dynamics of Cx32 HCs stimulated by a physiological $[Ca^{2+}]_c$ transient using combined imaging and patch-clamp techniques.
10. Experiments with R_m less than 1 GΩ (determined immediately after the whole-cell was achieved) should be discarded from the analysis as they can be indicative of previous Cx32 HC openings.
11. Fura-2 ratio was defined as $R(t) = F_{360}/F_{380}$, where F_{360} and F_{380} are emission intensities recorded at 520 nm during excitation at 360 and 380 nm, respectively.

Acknowledgments

This work was funded by grants awarded to Mario Bortolozzi by the French Telethon Foundation and by Theraphies for Inherited Neuropathies (TIN US) non-profit organization.

References

1. Delmar M et al (2018) Connexins and disease. Cold Spring Harb Perspect Biol 10(9)
2. Harris AL (2007) Connexin channel permeability to cytoplasmic molecules. Prog Biophys Mol Biol 94(1–2):120–143
3. Alexander DB, Goldberg GS (2003) Transfer of biologically important molecules between cells through gap junction channels. Curr Med Chem 10(19):2045–2058
4. Cotrina ML et al (1998) Connexins regulate calcium signaling by controlling ATP release. Proc Natl Acad Sci USA 95(26):15735–15740
5. Valiunas V (2013) Cyclic nucleotide permeability through unopposed connexin hemichannels. Front Pharmacol 4:75
6. Ye ZC et al (2003) Functional hemichannels in astrocytes: a novel mechanism of glutamate release. J Neurosci 23(9):3588–3596
7. Takeuchi H et al (2006) Tumor necrosis factor-alpha induces neurotoxicity via glutamate release from hemichannels of activated microglia in an autocrine manner. J Biol Chem 281(30):21362–21368
8. Bruzzone S et al (2001) Connexin 43 hemi channels mediate Ca2+–regulated transmembrane NAD+ fluxes in intact cells. FASEB J 15(1):10–12
9. Zocchi E et al (2001) Paracrinally stimulated expansion of early human hemopoietic progenitors by stroma-generated cyclic ADP-ribose. FASEB J 15(9):1610–1612
10. Rana S, Dringen R (2007) Gap junction hemichannel-mediated release of glutathione from cultured rat astrocytes. Neurosci Lett 415(1):45–48
11. Orellana JA, Stehberg J (2014) Hemichannels: new roles in astroglial function. Front Physiol 5:193
12. Meunier C et al (2017) Contribution of Astroglial Cx43 Hemichannels to the modulation of glutamatergic currents by D-serine in the mouse prefrontal cortex. J Neurosci 37(37): 9064–9075
13. Cherian PP et al (2005) Mechanical strain opens connexin 43 hemichannels in osteocytes: a novel mechanism for the release of prostaglandin. Mol Biol Cell 16(7):3100–3106
14. Burra S, Jiang JX (2009) Connexin 43 hemichannel opening associated with prostaglandin E(2) release is adaptively regulated by mechanical stimulation. Commun Integr Biol 2(3): 239–240
15. Contreras JE et al (2003) Gating and regulation of connexin 43 (Cx43) hemichannels. Proc Natl Acad Sci USA 100(20): 11388–11393
16. Valiunas V, Weingart R (2000) Electrical properties of gap junction hemichannels identified in transfected HeLa cells. Pflugers Arch 440(3):366–379
17. Gomez-Hernandez JM et al (2003) Molecular basis of calcium regulation in connexin-32 hemichannels. Proc Natl Acad Sci USA 100(26):16030–16035
18. Lopez W et al (2016) Mechanism of gating by calcium in connexin hemichannels. Proc Natl Acad Sci USA 113:E7986
19. Carrer A et al (2018) Cx32 hemichannel opening by cytosolic Ca2+ is inhibited by the R220X mutation that causes Charcot-Marie-tooth disease. Hum Mol Genet 27(1):80–94
20. De Vuyst E et al (2006) Intracellular calcium changes trigger connexin 32 hemichannel opening. EMBO J 25(1):34–44
21. De Vuyst E et al (2009) Ca(2+) regulation of connexin 43 hemichannels in C6 glioma and glial cells. Cell Calcium 46(3):176–187
22. Kleopa KA, Abrams CK, Scherer SS (2012) How do mutations in GJB1 cause X-linked Charcot-Marie-tooth disease? Brain Res 1487: 198–205
23. Meier C et al (2004) Connexin32-containing gap junctions in Schwann cells at the internodal zone of partial myelin compaction and in Schmidt-Lanterman incisures. J Neurosci 24(13):3186–3198
24. Balice-Gordon RJ, Bone LJ, Scherer SS (1998) Functional gap junctions in the schwann cell myelin sheath. J Cell Biol 142(4):1095–1104
25. Nualart-Marti A et al (2013) Role of connexin 32 hemichannels in the release of ATP from peripheral nerves. Glia 61(12):1976–1989
26. Lev-Ram V, Ellisman MH (1995) Axonal activation-induced calcium transients in myelinating Schwann cells, sources, and mechanisms. J Neurosci 15(4):2628–2637
27. Mayer C et al (1997) Intracellular calcium transients mediated by P2 receptors in the paranodal Schwann cell region of myelinated rat spinal root axons. Neurosci Lett 224(1):49–52
28. Stevens B, Fields RD (2000) Response of Schwann cells to action potentials in development. Science 287(5461):2267–2271
29. Ino D et al (2015) Neuronal regulation of Schwann cell mitochondrial Ca(2+) signaling during myelination. Cell Rep 12(12): 1951–1959

INDEX

Fabio Mammano and Mauricio Retamal (eds.), *Connexin Hemichannels: Methods and Protocols*, Methods in Molecular Biology, vol. 2801, https://doi.org/10.1007/978-1-0716-3842-2,